Praise for *The Neanderthals Rediscovered*

'[Written] with confidence and verve... strikes an excellent balance between broad popular appeal and satisfyingly rich content'

Society for American Archaeology

'Has the fresh charm of treating human evolution as a curious story that leads to the Neanderthals, rather than as a moral tale that rises ever upward and inevitably to us godlike moderns'

Harper's

'Excellent and absorbing'

Current World Archaeology

'A nuanced and sympathetic perspective on these fascinating people'

Brian Fagan, Professor Emeritus of Archaeology, University of California, Santa Barbara

'Fresh, well-informed and highly recommended... The inspired pairing of a Palaeolithic expert and a historian of science makes for compelling reading'

Paul Pettitt, Professor of Archaeology, Durham University

About the Authors

DIMITRA PAPAGIANNI trained as a Palaeolithic and stone tool specialist at Cambridge and wrote her PhD on the Neanderthal sites of northwestern Greece.

MICHAEL A. MORSE has a PhD in the history of science, specializing in the history of British archaeology. He is the author of *How the Celts Came to Britain*

T0026488

The Neanderthals Rediscovered

THIRD EDITION

Dimitra Papagianni
& Michael A. Morse

With 77 illustrations

For Eleana, Yanni and Vasili

On the front cover: Reconstruction of a Neanderthal based on the
La Chapelle-aux-Saints fossils. Reconstruction by Elisabeth Daynès
of the Daynès Studio, Paris, France. Photo S. Entressangle/E. Daynès/
Science Photo Library

First published in the United Kingdom in 2013 by
Thames & Hudson Ltd, 181A High Holborn, London WC1V 7QX

First published in the United States of America in 2013 by
Thames & Hudson Inc., 500 Fifth Avenue, New York, New York 10110

Revised and updated edition 2022
Reprinted 2023

The Neanderthals Rediscovered © 2013, 2015 and 2022
Thames & Hudson Ltd, London

Text © 2013, 2015 and 2022 Dimitra Papagianni and Michael A. Morse

Typeset by Mark Bracey

British Library Cataloguing-in-Publication Data
A catalogue record for this book is available from the British Library

Library of Congress Control Number 2021944313

ISBN 978-0-500-29640-0

Printed and bound in the UK by CPI (UK) Ltd

MIX
Paper | Supporting
responsible forestry
FSC www.fsc.org **FSC® C171272**

Be the first to know about our new releases,
exclusive content and author events by visiting
thamesandhudson.com
thamesandhudsonusa.com
thamesandhudson.com.au

CONTENTS

Timeline

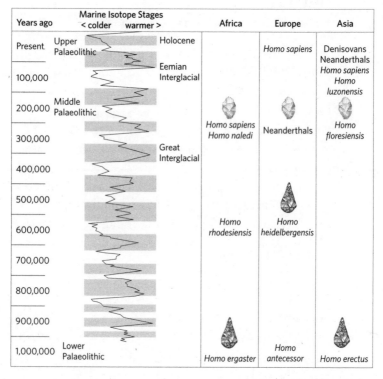

Years ago	Marine Isotope Stages < colder warmer >		Africa	Europe	Asia
Present	Upper Palaeolithic	Holocene		Homo sapiens	Denisovans Neanderthals
100,000		Eemian Interglacial			Homo sapiens Homo luzonensis
200,000	Middle Palaeolithic		Homo sapiens Homo naledi	Neanderthals	Homo floresiensis
300,000					
400,000		Great Interglacial			
500,000				Homo heidelbergensis	
600,000			Homo rhodesiensis		
700,000					
800,000					
900,000					
1,000,000	Lower Palaeolithic		Homo ergaster	Homo antecessor	Homo erectus

Handaxe   Levallois

Simplified timeline of the past million years, correlating climate changes (indicated by the Marine Isotope Stages, with the dark bands signifying warm stages) with different species of human and their stone tool technologies in Africa, Europe and Asia.

Preface

The rise and fall of the Neanderthals is one of the greatest stories from the prehistoric past. After more than 150 years of archaeological discovery, the Neanderthals are the best-known human species other than ourselves. They are closely related to us, yet distinct in both body and mind. And they coexisted with us quite recently: for the last Neanderthals, the earliest modern human art in Europe may have been thousands of years old already. For all these reasons, we have long wanted to share our fascination with them. It was just a matter of timing.

To appreciate the legacy of the Neanderthals, it is necessary to think in two different time scales. On the one hand, they lived deep in the past, and the story of their migrations and technological developments takes place over hundreds of thousands of years. Meanwhile, our understanding of them changes over the course of months, as modern technology enables us to answer questions that we could not even imagine asking just a few years ago.

Our challenge is that book editions have a lifetime between these two scales. Luckily, we have been helped out by odd coincidences. This project shares a birthdate with our twin sons. On the same day in 2007 that we welcomed them into the world, Thames & Hudson offered us a book contract. This bizarre moment, when the various time scales all came together on one particular date in the calendar, was fortunate for us (although at the time, this really seemed not to be the case!). With two newborns in the house, we had to postpone the writing, and we delayed the project long enough that we were able to include some crucial new discoveries that became central to the story. The most significant development was the publication of the first Neanderthal genome in 2010, which contained the dramatic news that for non-Africans living today, a small percentage of their DNA is inherited from the Neanderthals. Later, it emerged that Africans, too, carry Neanderthal DNA, albeit in an even smaller percentage.

By the time this book was first published – in 2013 – we still did not know the significance of that tiny percentage of Neanderthal DNA in modern people, other than what it meant for the Neanderthals' extinction. Once again, we experienced a strange coincidence of time scales. We were giving a book talk to a group of medical students in January of 2014 when a hand went up from the back of the room. The question was a natural one for someone who may have just heard that she carries a little bit of Neanderthal DNA in her. She asked what it was coding for – what, if anything, it was doing for our development and health. The first attempt to answer that question had, funnily enough, been published in *Nature* that very morning.

A revised and updated paperback edition came out in 2015, and we were able to include much of the new information then available. That year marked the beginning of an enormous increase in activity in the field of palaeogenetics. Before 2015, the number of published ancient gene sequences was in the single digits. By the end of that year, it was in the hundreds. Since 2018, more than a thousand ancient genomes are published every year, although most of them post-date the Neanderthal times, and we have learned much about Neanderthal DNA found in modern genomes. There have also been many discoveries related to the Denisovans, who were briefly mentioned in the first edition and now have their own section. All this new information gives us the opportunity to add entirely new dimensions to our understanding of the Neanderthal legacy.

Our original motivation for writing this book was our conviction that the Neanderthal story is a compelling one – one that has not previously been told in its entire dramatic arc from origins to expansion to demise. Most books with the word 'Neanderthal' in the title seem to collapse the species into an ahistorical list of features and important sites before morphing into a book on the species that replaced them. We wanted to write a book on the Neanderthals that does not dwell too much on the false turns in the long history of research and does not get easily distracted by the entry of *Homo sapiens* onto the scene. In short, we envisioned a book on the Neanderthals that is fairly exclusively about the Neanderthals.

The need for a book like this became apparent when Dimitra started teaching a course called The Neanderthal World in various university continuing education programmes. Her course was unusual in that it looked at human evolution from the Neanderthal point of view, and she discovered that it was immensely popular. Students kept asking her to recommend a book, and she realized there was no good single source on the Neanderthals. The most comprehensive book was *In Search of the Neanderthals* by Chris Stringer and Clive Gamble. Published back in 1993, this was becoming out of date and did not have anything on the exciting new discoveries. While there are many good books to recommend that touch on the topic, we wanted to bring all the threads of the Neanderthal story together in one place.

It is impossible to write about a million years of prehistory, including sites from western Europe to central Asia, and about fields as disparate as biological anthropology, genetics, geology and archaeology, without tremendous help. We are indebted to our many friends and colleagues who have helped guide our thinking. This book could not have been written without them. At the same time, we have presented the Neanderthal story through the filter of our own idiosyncratic choices, and responsibility for those, and indeed any mistakes, is ours alone.

We would particularly like to thank Mircea Anghelinu, Nick Ashton, Jill Cook, Clive Gamble, Ivor Karavanić, Mark Lamster, Paul Pettitt, Nellie Phoca-Cosmetatou, Matt Pope, Wil Roebroeks, Antonio Rosas, Katharine Scott, Chris Stringer, Aaron Stutz, Carolyn Szmidt and Carole Watkin.

Our original editor, Colin Ridler at Thames & Hudson, played a very large role in shaping the book, and we owe him a huge debt of gratitude. Sincere thanks are also due to Kim Richardson, Louise Thomas, Sarah Vernon-Hunt, Aaron Hayden, Terry Martin, Celia Falconer and Steve Russell.

For this edition, we would like to thank our new editor, Ben Hayes, for his help and encouragement. Our thanks also to Isabella Luta and the whole team at Thames & Hudson. We are grateful to Evan Eichler and David Reich for helping us understand ancient DNA.

As our sons, and with them this project, now enter their teenage years, it is interesting to reflect on how the popular image of the Neanderthals has evolved over the course of three editions. We are pleased to report that it has been a long time since anyone has joked to us that we should include [insert name of loathed politician or disliked colleague here] in a book about the Neanderthals. At the start of the project, this had been a constant refrain. Either the joke has grown old or, knowing that some Neanderthal blood flows through our veins, it just isn't funny anymore.

A Long Underestimated Type of Human

From our perspective as *Homo sapiens*, it can be tempting to look back on human evolution with a sense of triumph and destiny. With our large brains, crafty hands, agile legs and complex social networks, it was surely inevitable that when the conditions became favourable, we would assume our rightful place at the very pinnacle of nature. But perhaps it is a little too easy to claim this now that we have no serious rivals, and certainly no human ones.

We are only one variety, 'modern humans', of up to a dozen human species that have shared our world in the last two million years or so. And our direct ancestors were not always in the leading position. If we could rewind the clock to the last time the world was as warm and stable as it is today – some 120,000 mostly icy and climatically erratic years ago – we would probably not feel as privileged as we do now. Back then, the Neanderthals – a species much like us, only stronger – were on the march. With the capability to survive cold climates, this rival species might have been the prime candidate to populate the whole Earth from its origin point in Europe and push other forms of human to extinction. Over some tens of thousands of years from that point, the weather deteriorated, and enormous ice sheets advanced in the northern latitudes. Somehow, the particular human species native to Europe managed to survive these hardships, while our ancestors faced warmer challenges in Africa and, later, tropical Asia.

Fast-forward to about 45,000 years ago, when modern humans first ventured into a Europe half-covered by glaciers. There they encountered a form of human that, unlike themselves, had already lived through several cold episodes. Yet it was the warm-adapted *Homo sapiens* that survived this most recent glacial cycle. The cold-adapted Neanderthals failed, dying out in the millennia after our ancestors arrived in their homeland.

The species equivalent of our first cousins, the Neanderthals, were unmistakably different from modern humans, with large, barrel-shaped chests, stocky and muscular bodies, broad noses and chins that did not protrude. They shared many of our behaviours, but their development seemed to trail ours in some key areas, such as our capacity for symbolic expression.

Even their world was alien to ours. The open, steppe-like landscapes that prevailed in Europe in their heyday enjoyed good exposure to sunlight and had rich vegetation that sustained sizeable populations of mammoth, bison, deer and horse – species essential to the Neanderthal diet. Neanderthal landscapes were unlike the barren steppes of present-day Eurasia or, indeed, any landscapes anywhere in today's world.

The Neanderthals are the only predominantly European species in the human family tree, and this alone has fostered an enduring fascination with them. Europe has a much longer history of archaeological investigation and research, and therefore a richer record of its distant past, than any other part of the world. We have known them longer than any other extinct form of human.

And because the Neanderthals are a relatively recent species, their fossilized bones, the remains of their everyday lives and even their DNA are well preserved. The Neanderthals are not just one of our closest relatives. They are also the ones we know best. With Neanderthals, we are as close as we will ever get to another species from the human evolutionary past.

Like so many people who are different from us, the Neanderthals are now known mainly for the use of their name as a pejorative. In this book, we do not pretend to be able to correct this popular usage, but we do hope to restore some dignity to those we replaced.

The Neanderthals have always been a little too close for comfort for the modern Western world. Their name conjures up images of muscular but dim-witted cavemen who relied on force over cunning. When a *New York Times* arts critic wrote about 'Neanderthal TV' in 2005, he was referring not to a documentary on human evolution, but to programmes that feature 'deeply flawed' male characters with 'antisocial tendencies'. The name can also be a source of humour. A noughties rock band called The Neanderthals dressed in animal skins and sang simplistic songs about girls. We think this is all unfair.

Too often the Neanderthal story is told simply as a backdrop to our own, a reflection of the same obsession with our own species found in our creation stories. Yet there is a gentle counter-current to this *sapiens*-centrism, a growing sense of collective guilt that appears in the most unlikely places. We once came across a bottle cap that, for no clear reason, declared: 'The brain of Neanderthal man was larger than that of modern man.' It was a reminder that, of all the large mammal species whose extinction we have witnessed in recent millennia, at least one seems to have been our equal in brain capacity, if not in humanity.

In recent years new research has pulled the Neanderthals much closer to us. Not only did they have brains as large as ours (though their skulls had a different, flatter shape), they also buried their dead, cared for the disabled, hunted animals in their prime, used a form of spoken language and even lived in some of the same places as the modern humans who were, broadly speaking, their contemporaries. They could not have survived, even in warmer times, had they not mastered fire and worn clothes. Though they relied heavily on meat, they consumed seeds and plants, including herbs, and could fish and harvest sea food. These are all behaviours that at some point were thought to be exclusive to ourselves.

A golden age of research

The pace of progress in our understanding of the Neanderthals keeps accelerating. Thanks to some breathtaking recent discoveries and scientific advances, we can now examine the story of the Neanderthals in greater depth than was previously thought possible. This is truly the golden age of Neanderthal research, making it the perfect time to see how all the strands of evidence fit together into a narrative of the rise and fall of a long underestimated type of human.

When we started studying the Neanderthals in the early 1990s as graduate students, much of what is now our current knowledge was the subject of heated debate. Most of this debate revolved around the Neanderthals' role in our own story. In those days, for example, it was not yet clear whether the Neanderthals and modern humans ever shared the same continent at the same time. (We now know they did.)

Archaeologists were then still absorbing the implications of a genetic study that supported the so-called Out of Africa theory, that all living modern humans can be traced back genetically to a single woman (or a small group of women) in sub-Saharan Africa. The entry of genetic evidence into archaeology exposed sharp divisions in the discipline. Although the Out of Africa theory was about the evolutionary trajectories of all modern human populations, Europe and the Neanderthals were a key part of the debate. The dominant question in Neanderthal research at the time was whether they were part of our evolutionary ancestry or whether they had been replaced, either after a hiatus or by being out-competed – perhaps even killed off – by modern humans who had originated in Africa and migrated into Europe.

In 1993 Chris Stringer and Clive Gamble published their land-mark volume, *In Search of the Neanderthals*, which put forward the case that modern humans replaced, rather than evolved from, the Neanderthals. Little did we realize that this was just the beginning of a wave of new insights and evidence about our evolutionary kin, and that the question of our replacement of the Neanderthals would re-emerge forcibly with further genetic studies.

Since then, there have been several major news stories every year on the Neanderthals. Atapuerca, a cave network in Spain, has produced the remains of around thirty individual 'proto-Neanderthals', an astonishing total if we consider that the number of Neanderthal individuals in the fossil record is only a few hundred. In 2007 a discovery from a different part of Atapuerca pushed definitive evidence for the earliest occupation of Europe, possibly by a Neanderthal predecessor species, over the 1-million-year mark for the first time. And more recently, a team in Germany led by Svante Pääbo (see plate 15) has made staggering claims based on the identification of Neanderthal DNA, leading to a whole new subdiscipline. It can be hard at times to make sense of this flood of information.

The number of scientific disciplines contributing to Neanderthal research in recent years has multiplied. Our knowledge has been boosted by a variety of specialists: geologists who drill sediments from the deep ocean floor and take ice cores from glaciers covering Greenland and Antarctica; archaeologists who sieve and sort through every ounce of dirt from an excavation looking for anything from seeds to rodent teeth; geneticists who put on sterile clothing to drill Neanderthal bones in the hope of finding bits that have neither fossilized nor been contaminated by flakes of their own skin in order to extract ancient DNA; and, of course, those archaeology students who kneel down with a trowel and a brush to excavate squares of earth, layer by layer, carefully recording the location of every stone tool or bone they might find (and hoping they aren't embarrassed by confusing the two).

We now know more than we ever thought we could about the Neanderthals and their world. They have emerged as more accomplished in their everyday lives and more complex in their social behaviour than we imagined. This makes it all the more puzzling why this separate form of human, nearly as advanced biologically and culturally as ourselves, became extinct. The fate of the Neanderthals has been a mystery for more than 150 years.

The valley of the new man

But before we consider the Neanderthals' fate, let us explore their discovery. There is a satisfying irony in the fact that we discovered our former human rivals in the course of supplying the energy and materials for global domination. It is no exaggeration to say that our knowledge of the Neanderthals was an unexpected by-product of industrial mining in the 19th century. As engineers were digging ever deeper for minerals, evidence was fast accumulating about the Earth's past.

'There will be very soon little left to discover,' was how *The Times* of London described the science of geology in 1863. The occasion was the British Association for the Advancement of Science's annual summer meeting in Newcastle. The conference was enormous, with scores of mineralogists, geologists, chemists and others from dozens of burgeoning scientific disciplines. It was here that a little-known professor called William King became the first person to use fossil evidence to name an extinct species closely related to our own.

The conference's keynote speaker was Sir William Armstrong, a local industrialist and engineer. In celebrating the fast pace of discovery, he noted that in short order Charles Darwin had finished *On the Origin of Species*, John Speke and James Grant had found the source of the Nile and Charles Lyell had published *Geological*

Sir William Armstrong celebrates the rapid progress of science in his keynote address to the 1863 meeting of the British Association for the Advancement of Science in Newcastle-upon-Tyne, as depicted in an engraving from the *Illustrated London News*. Later in the conference, the Neanderthals received their scientific name, *Homo neanderthalensis*.

Evidences of the Antiquity of Man, which extended human history deep into the past.

One of Lyell's protégés, Professor King had come to the meeting from Queen's College, Galway, in Ireland to present a short paper on a recent discovery from a lime quarry near Düsseldorf, in what was then Prussia. Some sixteen years earlier he had been forced out of his job as curator of the Newcastle Museum. Now he hoped to return in triumph.

In his paper King discussed human-like bones that had turned up in a cave (Feldhofer) in the Neander Valley seven years earlier. The collection included ribs, arm and leg bones and the top of a skull that featured a protruding crest above the eyes. In a departure from the prevailing view that the bones belonged to a deformed member of our own species, King argued that they dated to the

Professor William King, of Queen's College, Galway, Ireland, who suggested the name 'Homo Neanderthalensis'.

Drawing of the original Neanderthal skull from *Man's Place in Nature* by Thomas Huxley.

glacial period and were closer to the chimpanzee than to any modern human. We now know that King was wrong on this last point, for the individuals from the Neander Valley were much closer to modern humans than chimpanzees. But he was correct in his more startling conclusion, when he argued that they were fundamentally different from all living humans.

He proposed 'to distinguish the species by the name Homo Neanderthalensis' after the Neander Valley where they were discovered. In the process, he unwittingly immortalized a 17th-century psalmist, Joachim Neumann. Following the Philhellenic fashions of the 19th century, Neumann's last name (literally 'new man') was translated to the Greek 'Neander' and then attached to the valley ('thal' in German) where he penned his psalms. Thanks to King, members of the long-extinct species unearthed there have been known by the wonderfully ironic name of Neanderthals (people from the valley of the new man).

Before that day, the nearest known species to humans were living apes. Today there are more than twenty named species in our family tree since the split from the apes, and with more recent discoveries such as *Homo floresiensis* – the so-called 'Hobbit' – in Indonesia in 2003 and the Denisovans from Siberia whose genes were identified in 2010, it seems likely that this number will continue to increase. In fact, since the first edition of this book in 2013, we have welcomed *Homo naledi* (from South Africa) and *Homo luzonensis* (from Indonesia) to the human family.

Before the discovery in the Neander Valley in 1856, other Neanderthal remains – a small child's skull from Engis, Belgium, in 1829, and an adult skull from Gibraltar in 1848 – had been unearthed, but they had not been recognized as a distinct species. Additional bones appeared in the ensuing decades, notably from Spy, Belgium, in the 1880s. Yet the Neanderthals lingered on the margins of the human evolutionary story, even as other species, such as *Homo erectus* in its guises as Java Man (discovered at Trinil, Indonesia, in the 1890s) and Peking Man (discovered at Zhoukoudian, China, in the 1920s and 1930s), quickly acquired the label of 'missing link' between ourselves and the apes.

Today, evidence of Neanderthal occupation extends from Wales all the way to Siberia. In this book, we trace the evolution of the Neanderthals in Europe, their expansion into Asia and their ultimate encounter with our own ancestors. Each successive chapter moves us closer to the present and covers smaller and smaller chunks of time. This is not because the distant past necessarily contained fewer noteworthy developments – though change seems to have occurred more slowly then – but because the evidence of more recent times is better preserved.

New questions

With all the recent advances in Neanderthal research, there are new, intriguing and sometimes unexpected questions. In addition, things we thought we knew, like scientific dates, are sometimes overturned in the light of new evidence. In each stage of Neanderthal evolution, and in every chapter in this book, there are questions to which we do not have definitive answers; but we can now at least make reasonable attempts to answer them. Part of the enjoyment for us, and we hope for you, is to use the available evidence to put together a plausible story.

We start around 1 million years ago with the first human entry into Europe, which would ultimately become the Neanderthal homeland. Who were Europe's first colonizers, and why did it take so long for them to get there, when humans had gone from Africa to southeast Asia almost a million years earlier? (Intriguingly, this pattern is mirrored much more recently by modern humans, who also reached southeast Asia before Europe.) Did they arrive in Europe from Africa or Asia? How successful was this first wave?

While Europe has been inhabited by human species since around 1 million years ago, it was only by 500,000 years ago that incipient Neanderthal traits first appeared. Was this a case of local evolution or did a new group arrive? The Europeans at that time used the same primary stone tool – the handaxe – that was favoured by people throughout Africa and much of Asia. What led the Europeans to take a different evolutionary path from their neighbours when they had so much in common?

Neanderthals took on a distinctive form by 250,000 years ago. As their bones became unmistakably Neanderthal, their stone tools changed, and in a sense their technology improved. Instead of the large, unwieldy handaxe, they started using lighter, pocket-sized stone tools with long, sharp edges. The biggest mystery about this new kind of cutting tool is why it appeared almost simultaneously

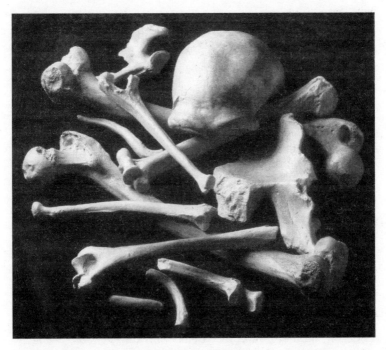

The bones discovered in 1856 in Feldhofer Cave in the Neander Valley near Düsseldorf, Germany, that led to the first identification of an extinct human species.

in at least three locations, Europe, western Asia and southern Africa. Does this indicate some form of contact between early Neanderthals in Europe and the ancestors of modern humans in Africa, or are lightweight, versatile stone tools an inevitable outcome of larger brains? This begs the question of why Europeans and Africans, apparently in parallel, were both evolving larger brains.

A sustained warm period that began some 130,000 years before the present enabled Neanderthals and modern humans to expand simultaneously from their separate homelands, ending the era of isolation in which they had evolved into separate species. (We follow the convention of calling them separate species on the grounds of the differences in the shapes of their bones, and not because of any opinion as to their ability to reproduce successfully together.) As the two human lines spread into Asia, it was not yet clear which of them was better positioned to take advantage of the opportunity to increase their population size. At the time there were perhaps five or six distinct types of human. To us, the world feels small. With modern transportation, we can reach almost any destination within twenty-four hours. How did our little planet come to support so many different forms of humanity?

Clutching very similar stone tools, *Homo neanderthalensis* and *Homo sapiens* were the prime contenders to dominate the Earth, coming out of Europe and Africa, respectively. It is unclear which human species were in Asia at this time: whether *Homo erectus* still survived in east Asia, or if the recently discovered dwarf human species, *Homo floresiensis* – possibly an even more archaic species – was then living on the island of Flores in modern Indonesia. It is likely that *Homo luzonensis* was still living on the Indonesian island of Luzon. Another group, called Denisovans and likely also a separate species, seems to have been widespread in Asia. How close was the Earth to becoming a Neanderthal world? The safe money was on the Neanderthals emerging as masters of the planet. They were among the strongest of these species. They had survived the kind of cold conditions that were making life increasingly difficult for everybody. And their brains were large. Modern humans were contenders, but were physically much weaker than the Neanderthals and had no chance in straight combat. The only tiny advantage we know that *Homo sapiens* had was that it was extensively producing jewelry with little beads and shells. The significance of this would later become clear.

Modern humans and Neanderthals both reached the Near East some time after 130,000 years ago. This was one of the most interesting moments in prehistory, as two human species roamed over the same territory. There is much we do not know about that epoch, starting with the question of whether the two species were there at the same time, whether they encountered each other and, indeed, if they interbred. What we do know is that this is close to the time when modern humans began to establish themselves as masters of the planet. But Neanderthals were also increasingly successful, reaching deeper into Asia.

Everything changed for the Neanderthals by around 40,000 years ago, when *Homo sapiens* penetrated deep into their European homeland. This marked the end for a proud species of human. But how did they fail? Did they go quietly, retreating and dwindling away? Were they hostile or did they welcome the new arrivals? And how did we emerge victorious? To what extent did interbreeding play a role?

Alternative futures

We are now the sole survivors in what used to be a diverse human world. It has been left to science-fiction writers to imagine an alternative future in which pockets of Neanderthals live on in places such as Tajikistan, northern California and Swindon, UK. We cannot help but wonder: were it not for our own advances, would the Neanderthals have colonized the rest of the planet, replacing holdovers of more archaic species? Would they have accomplished all the things that modern humans have done – agriculture and architecture, states and warfare, science and psychoanalysis?

There is a comfort in imagining the Neanderthals as a sort of back-up plan for human domination of the planet, had we not been around. On the other hand, when considering the extinction of our closest relatives, it is hard not to think of our own prospects. In a strange sense, the price we have paid for our knowledge of the Neanderthals – or at least for the civilization that has enabled us to learn about the deep past – is the very stability that has got us this far. Our world is about to be shaken by another major change in climate. If we suffer a significant jump in warming, as we almost certainly will over the next few generations, or if we experience a new glaciation, which will surely come unless global warming somehow delays its arrival, our survival as a species may hang in the balance.

These grand thoughts have been a part of Neanderthal research from the beginning. Let us return to Sir William Armstrong's speech to the British Association in 1863, when he considered the impacts of the industry that had brought Newcastle to prominence in his day. The previous twenty-five years, Armstrong declared, had seen the emergence of the railway and transatlantic steam navigation to the point where life had become almost unthinkable without transportation driven by coal. Even at this early stage of industrialization, Armstrong foresaw that the prosperity of Newcastle in 1863 would not last for ever.

In an eerie echo of today's predictions about the end of the oil age, Armstrong calculated that Britain's coal reserves would run out by the year 2075, and he turned to the promise of solar energy as the likely guarantor of continued progress. As we edge closer to that date, it is not the end of energy supplies that worries us most, though that is indeed a great concern, but the effect that two hundred years of accelerating greenhouse gas emissions (of which coal burning is a leading contributor) will have on the global climate system. And we are still waiting for solar energy to live up to its promise.

Yet Armstrong was certainly confident about our prospects. Thanks to Lyell and his colleagues, the scientists of 1863 knew 'that the human race has existed on the earth in a barbarian state for a period exceeding the limit of historical record', dating back to a time when global temperatures were much colder. Humanity had survived extreme hardship before.

The Neanderthals also survived numerous episodes of extreme climate change, but they did not live to see the ice melt from northern

24

Europe at the start of the Holocene – the current warm period that began some 10,000 years ago. As we look at the rise and fall of our closest human cousins and at our own inheritance of a world without other human species, we hope to understand what the future may hold for us.

Will we go the way of the Neanderthals, who, after all, lived for hundreds of thousands of years in Europe as opposed to our few tens of thousands? Compared with what the Neanderthals experienced, the modern human story has only just begun.

The First Europeans

1 million to 600,000 years ago

For the first five or six million years since they split from the apes, our ancestors did not enter Europe. There have been only three or four ape-descendant species in our family tree – what we call hominins – to live there. The Neanderthals are the only species – out of around twenty identified in our lineage – that we are reasonably certain evolved there. From the point of view of deep prehistory, *Homo sapiens* are relative newcomers to the continent that was the Neanderthals' birthplace and long-term homeland. Yet we do not know how long the Neanderthals' ancestors inhabited Europe. Did they arrive more than a million years ago, with the first human occupation, or was it just 600,000 years ago, following a period of extreme cold that may have wiped out the European pioneers? As modern humans, we have a stake in this issue, because it has implications about the timing both of our evolutionary separation from the Neanderthals and of possible ancient gene flow between Europe and Africa.

What we do know is that by 1 million years ago parts of Europe were occupied by a human species. But there are many unanswered questions about this species, starting with what it should be called: *Homo erectus* (which was present in Asia in this period), *Homo antecessor* (a name proposed by a Spanish team which uncovered some of the oldest European fossils) or *Homo heidelbergensis* (the name of a later European species)? Whatever species it was, did it arrive directly from Africa or was it an offshoot of an established Asian *Homo erectus* population? Why did it lack the characteristic *Homo erectus* stone tool, the handaxe, which was common at that time in Africa and the Near East? Did it give rise to the Neanderthals, or did it go extinct due to increasingly bitter cold periods or, perhaps, competition with later arrivals?

None of these questions could have been asked before the mid-1990s. Up to then, there was little sign that humans had been in Europe before a mere 500,000 years ago. But conclusive evidence emerged from a handful of sites in Spain and Britain to push that date back by a significant margin. Two of these sites are located within a few hundred metres of each other at Atapuerca in northern Spain: Sima del Elefante ('Pit of the Elephant', named after a fossil elephant found there) and Gran Dolina ('Great Basin') are part of a cave complex that is remarkable for its unexpectedly ancient

Map showing key sites discussed in this chapter of some of the earliest human activity outside Africa.

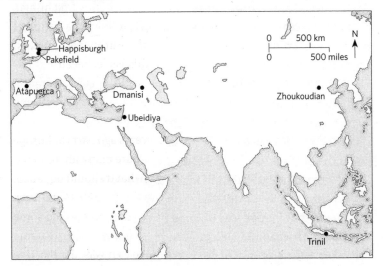

dates – the earliest being up to 1.2 million years ago – and large quantity of human remains.

In East Anglia in Britain there are two more sites, separated by a mere 56 kilometres (35 miles) of coastline. Both of these made headlines in the 2000s: in 2005 Pakefield was dated by the teeth of an extinct vole species to 700,000 years ago, earlier than archaeologists had thought that humans could have reached as far north as Britain, and in 2010 Happisburgh, lying even further north, was dated as slightly older by the same vole, along with other species, and by the magnetic orientation of fine-grained sediments.

In both the Spanish and British cases, there were fortuitous local conditions that led to the original preservation of the sites and to their recent exposure. This explains why in the two countries we find sites within a relatively small area. What can we infer about the population over the rest of Europe? The secure dates from Spain and Britain have led to a reassessment of a host of other sites in countries ranging all the way to Russia, where archaeologists had previously claimed to find stone tools from this early period but where the evidence was ambiguous. Even if all these older dates are confirmed, and there is plenty of debate, the evidence from the period 1 million to 600,000 years ago is thin and indicates a small population of European colonizers.

For the most part, the distribution of the early sites shows that the human population was still adapted to living in warm climates like its African ancestors. Most of the sites are found in southern Europe; even sites located as far north as Britain were in use mainly when the climate was very warm. The lone exception to this is Happisburgh, the excavators of which argue it was occupied during chillier times. In any case, these sites signal the end of the long-held belief that Europe was isolated from the first human expansion out of Africa until half a million years ago. In this chapter, we will explore the evidence of the first human occupation of Europe during the time before ice played a significant role in its climate, and consider what became of the initial settlement of Europe: did it ultimately fail, or did it succeed and lead to the evolution of the Neanderthals?

The first world traveller

The genus *Homo* evolved in Africa, and the ability to expand to more than one continent was one of its defining traits. The first human exodus from Africa occurred around 1.9 million years ago, and descendants of these early hominins seem to have inhabited the warmer parts of Asia to as late as 100,000 years ago. This is an extraordinarily long occupation, perhaps twenty times longer, for example, than our own species' presence in Asia.

The human ancestor that we tend to give credit for this first world conquest is *Homo erectus* ('upright man'), although some now argue that an earlier species related to *Homo habilis* ('handy man') may have been the first to leave Africa. (Evidence is so scattered that others consider at least the possibility that the genus *Homo* evolved in Asia and back-migrated into Africa.) All agree, however, that *erectus* was among the most successful of our ancestor species. The main features that separate *Homo erectus* from its predecessors are in its legs and pelvis, the most important part of the body for long-distance walking, and its brain size, which grew to about 75 per cent of ours.

What gave rise to these changes? The most popular explanation links them to the beginning of the current Ice Age about 2.5 million years ago. When we think of the Ice Age, we might think of desolate, glacier-filled landscapes. But geologists consider even the present day to be part of this Ice Age, which is characterized by long eras of cold climate (glacials), in which glaciers expanded, punctuated by shorter warm periods (interglacials), like the Holocene we currently enjoy. As the Ice Age progressed, the cold periods became increasingly extreme and long-lasting, and by about 650,000 years ago Europe suffered its first major glaciation.

When the Ice Age began, it brought drier conditions to Africa (forests giving way to savannahs) which in turn may have forced our ancestors to modify their plant-based diet, as their woodland habitat contracted, in favour of a more meat-heavy one that required greater long-distance travel to find food. Scavenging and possibly hunting for meat and fat required an enormous expansion in feeding range and led *Homo erectus* to follow migrating animals. These may be key factors in explaining why *erectus* was able to penetrate deep into Asia almost as soon as it first appeared.

The change to more intensive meat consumption had many advantages. Animal protein is digested more easily than plants and nuts and allowed for a reduction of the intestinal tract and commensurate expansion of the brain. The logic behind this, known as the expensive tissue hypothesis, developed by anthropologists Leslie Aiello and Peter Wheeler, is that the intestines require a large amount of energy, though not quite as much as that required by the brain, the most 'costly' organ in our bodies. Therefore a shorter intestinal tract would have freed up energy for a larger brain.

This evolutionary development may have triggered a positive feedback loop, where increased meat consumption enabled a larger brain to develop, which in turn led to an increased ability to obtain meat. Some researchers believe that our species continues to evolve towards improvements in the brain, while, ironically, we have advanced to the point where we know how to control our diet so that we can 'feed' our demanding brains without needing to eat meat at all.

The fossil record is fairly sparse for this period, especially outside of Africa. How much do we know about *Homo erectus*? With *Homo erectus*, we have either the most successful human species to date (measured by its longevity) or we have something of a generic term that conceals a more complicated story. In the nebulous world of early human evolution, where there are almost as many named human species as there are known fossil sites, *Homo erectus* is really the one solid point of reference. It is the only Asian hominin (other than ourselves and the Neanderthals) whose very existence as a separate species is not under serious dispute. *Homo erectus*'s privileged position in the human story has a lot to do with its extraordinary discovery.

At a time when the only known pre-*sapiens* fossils were Neanderthals, a Dutch anatomist called Eugène Dubois travelled to Indonesia and started digging in a quixotic bid to find a 'missing link'. In the early 1890s his teams began, incredibly, to succeed on the island of Java. At the site of Trinil he discovered a small, primitive brain case near a long leg bone and proposed the name *Pithecanthropus erectus* ('upright ape-man').

The great dinosaur hunter O. C. Marsh celebrated the discovery in the *American Journal of Science*, saying: 'He has proved to science the existence of a new prehistoric anthropoid form, not human indeed, but in size, brain power and erect posture, much nearer man than any animal hitherto discovered, living or extinct.' Marsh wrote these words more than thirty years after the naming of *Homo neanderthalensis*, but the Neanderthals were, to him, even less human than *erectus*. By the mid-20th century, as the humanity of the species Dubois discovered was becoming more apparent, the designation became *Homo erectus*.

Ironically, the leg bone that was so central to the naming of *erectus* may in fact be modern. Nevertheless, Dubois's concept of a transitional species that walked upright and ranged over Asia has been supported by discoveries of a number of individuals who are clearly similar to the first fossils he found.

The brain case that Dubois unearthed is now the type-fossil for *Homo erectus*, which means that it is the reference point for classifying later discoveries. Yet it is probably less than a million years old and is too fragmentary to be of great use in comparing it to other fossils. In other words, *Homo erectus* has the benefit of a vague definition, which in any case dates to perhaps a million years after the first African exodus.

The earliest hominin site outside Africa is Dmanisi in Georgia, which is dated to between 1.7 and 1.8 million years ago. The site lies under the ruins of a medieval hilltop village, and its discovery came about in the early 1980s when archaeologists found a fossilized rhinoceros tooth (long extinct in western Asia) within the medieval habitation. The five hominin individuals unearthed there have small brains for *Homo erectus*, and the variation in size and robustness among them leaves open the possibility that more than one species is represented. Many view the Dmanisi fossils as more closely related to an earlier African species, *Homo habilis*, and some researchers use the term *Homo georgicus*, arguing that the fossils are unique. The relationship between *Homo habilis*, *Homo erectus* and *erectus*'s more gracile African relative, *Homo ergaster*, is the subject of some debate, making classification less than straightforward. What we can safely say about Dmanisi is that the people there were probably a primitive version of *Homo erectus* or perhaps one of its immediate predecessors.

The stone tools from Dmanisi are similar to those found in African sites of the same age: simple flakes with sharp edges, produced in a few minutes by striking a piece of raw stone with a hard hammer. Like the humans at Dmanisi, some of the animal fossils were from species that had also emigrated recently from Africa, and it is possible that the hominins simply followed animal herds as they migrated and that the environment of Dmanisi was not very different from that of the sub-tropical areas of northern Africa or south Asia.

The next oldest Eurasian fossils are from Java and China and may date as far back as 1.5 or 1.8 million years ago, but most of the exact findspots are unknown. In China the earliest *erectus* fossils probably date to about a million years ago, though there are also controversial claims for finds as old as the earliest of the Java fossils. Ubeidiya in Israel dates to 1.5 million years ago.

The sites in Georgia and Israel show that early humans were at the doorstep of Europe close to the time of *Homo erectus*'s first appearance. But *erectus* was adapted to tropical and sub-tropical environments. Europe lies at a high latitude, and its harsh winters and short growing seasons for vegetation may have made it unattractive to the early pioneers out of Africa.

Homo erectus's signature stone tool, which appeared 1.76 million years ago in east Africa and in Israel afterwards, is the handaxe. The handaxe is a large, multi-purpose cutting device, knapped into shape on both sides, and which in most cases fits comfortably in the hand. It was useful for virtually all aspects of butchering large

Three views of a handaxe found at Abbeville, France. In the 19th century, the discovery at Abbeville of handaxes such as this one, in association with bones of extinct animals, led to the recognition of the great age of human existence on Earth.

game (whether hunted or scavenged), including separating flesh from bone, skinning to make hides and possibly also killing the animals. It may have had other uses as well.

The importance of this tool in the history of science, let alone for human evolution, is generally underappreciated. One of the greatest achievements of modern science is the establishment of human antiquity, which can be traced to the discovery of handaxes in the Somme River gravels at Abbeville, France, in the 1830s and 1840s. Jacques Boucher de Perthes unearthed many handaxes in a layer with bones of animals such as elephants, rhinos, hippos and tigers, which belonged to an earlier geological epoch. These were among the first artifacts that showed that humans have existed on the Earth for more than just a few thousand years, which was the time-frame that had been generally accepted before this discovery.

For the wider scientific community, the presence of these tools in ancient strata – and therefore the great antiquity of their human makers – was confirmed after British geologists Joseph Prestwich and John Evans visited the gravels in April 1859. One of the key sites they visited was at Saint-Acheul, and today the technical term for these handaxes is Acheulian. These events were a major factor in the 1863 publication of Charles Lyell's *Geological Evidences of the Antiquity of Man*, which was widely praised at the British Association meeting where William King named the Neanderthals, as we discussed in Chapter 1.

Handaxes continue to have an enormous impact on our knowledge of prehistory. For example, at Happisburgh on Britain's Norfolk coast, a cluster of five separate sites from different periods – including the very ancient one discussed in this chapter – came to light thanks to the discovery of a handaxe in tidal waters.

Judging by its longevity, the handaxe is the greatest-ever human invention. From its first appearance, humans continued to use this tool until around 40,000 years ago. This means that the tradition of making handaxes lasted far longer than its original inventors and was practised by a number of *erectus*'s successor species, probably including our own. Although early *Homo erectus* handaxes were common in east Africa and Asia west of the Himalayas, the European pioneers did not use them until 500,000 years ago. This suggests that the east Asian and European populations may have been isolated from later innovations arising in Africa. In east Asia, however, we can attribute the difference to raw materials, as bamboo was probably used as a more convenient alternative. In Europe, the handaxe simply arrived late.

When humans did arrive in Europe, what route did they take? On current evidence, it took more than half a million years for humans to reach the continent following their initial occupation of Dmanisi. The earliest known site in Europe is all the way on the other side, at Atapuerca, in Spain. But Spain is not the most likely gateway to the continent, because there was no land bridge to north Africa, and there is no evidence that any hominin crossed the sea at such an early date. This leaves open the possibility that there are other ancient sites in central or eastern Europe that have not yet been discovered. We are currently experiencing something of a golden age of discovery of sites from this period, and the textbooks may need to be rewritten yet again.

Explaining the Ice Age cycles

Once *Homo erectus* and its successor species became widespread throughout Eurasia, the increasing frequency of glacial cycles became a major driver of evolutionary changes, culminating in the appearance and subsequent extinction of the Neanderthals. We therefore divert our attention briefly to the phenomenon of Ice Age cycles, which played a prominent role in the Neanderthals' changing fortunes.

Over the past million years, glaciers have expanded and retreated in Europe perhaps ten separate times. Some of these glaciations have been severe, others less so. The alternation between cold glacial periods and warm interglacials has itself not been uniform, and there have been major changes in climate that have been as short as a human lifetime. Since the existence of these glacial cycles was discovered in the 19th century, their causes have confounded researchers. However, a combination of astronomical observations, deep sea drilling and complex computer modelling has moved us closer to understanding this pattern.

These climate fluctuations, which have become more extreme in the past million years, started when the Earth entered the current Ice Age about 2.5 million years ago thanks to a number of factors: the drift of the continents towards higher latitudes; the emergence of the isthmus of Panama to join North and South America, diverting warm equatorial waters away from the poles; and the uplift of the Himalayas and Tibetan plateau. These provided favourable conditions for the expansion of ice sheets while also reducing the amount of the greenhouse gas CO_2 in the atmosphere. Once the Earth cooled to a critical point, astronomical cycles, which affect the amount of solar radiation the Earth receives, began to have an influence on global temperature, and glaciations became more severe.

During the Ice Age, the climate is influenced by three key astronomical cycles, called the Milankovitch cycles, after Milutin Milanković, the mathematician who first proposed them in the early 20th century. The effect of the Milankovitch cycles is that the amount of sunshine that reaches different parts of the Earth's surface does not remain the same over time. For Milanković, a key factor in glaciations is the amount of solar heat that falls in the high latitudes of the northern hemisphere during the summer. If the summers are cool, he reasoned, then the winter snows do not melt entirely, and each year there is a build-up of snow and ice which extends gradually southwards.

Everyone is familiar with two basic cycles that affect sunshine penetration in similar ways to the Milankovitch cycles: the day/night rotation of the Earth and summer/winter seasonality. Seasonality is caused by the tilt of the Earth's axis which, over a year, alternates exposure of the northern and southern hemispheres to more direct sunlight. The three Milankovitch cycles operate over much longer intervals, but with the same basic principle.

The dominant Milankovitch cycle has to do with the shape of the Earth's orbit around the Sun. The Earth's orbit is not a perfect circle, but is in the shape of an ellipse, with the Sun set off the centre. This shape varies over a cycle of about 100,000 years, as the orbit oscillates between a near-perfect circle and an elongated egg shape. During the millennia when the orbit is egg shaped, the Earth travels further away from the Sun than when the orbit is nearly circular. This can lead to an accumulation of ice if the Earth receives less sunlight at key points in the calendar.

Within this major cycle, there are two shorter cycles. Every 42,000 years the tilt of the Earth's axis of rotation fluctuates between 22.1 and 24.5 degrees, affecting the intensity of winters and summers. And there is a 21,000-year cycle, set by the way the Earth 'wobbles' as it spins, affecting the temperature swing between seasons. If, as at present, winter in the northern hemisphere begins when the Earth is closest to the Sun, winters are relatively mild and summers are temperate. But in the other half of the 21,000-year cycle, the Earth will be furthest away from the Sun at the onset of winter in the northern latitudes, and the differences between seasons will be more pronounced, with harsher winters and hotter summers.

These three cycles run simultaneously, pushing the Earth's climate in one direction or the other. Most of the time, these three cycles push and pull in different directions, and their combined effect helps to explain why each major cold or warm phase is not uniform, but is interspersed with shorter warm or cold episodes.

It was not until the 1970s that Milanković's theory was confirmed thanks to two sources: cores taken from deep sea sediments and ice cores taken from Greenland and Antarctica. The ocean floor drilling started in the 1950s, and the ice cap drilling started in the 1970s, with research in both areas continuing today. The more recent ice core research has had an enormous impact on reconstructions of past climates because the ice cores have higher resolution, which means that they record shorter-term events, and so they have provided startling evidence for just how rapidly climate change has occurred in the past. For example, some warming events in the past 100,000 years seem to have happened over just a few decades.

In both types of cores, researchers can 'read' climate history from the ratio of regular oxygen atoms (^{16}O), which are made up of eight protons and eight neutrons, to a heavier oxygen isotope (^{18}O), which contains eight protons and ten neutrons. During glacial conditions, more of the regular (lighter) oxygen is stored in the ice caps, leaving a higher ratio of ^{18}O/^{16}O in the ocean and atmosphere. This ratio is then preserved in the fossilized shells of micro-organisms at the bottom of the ocean and in tiny air pockets in the ice cores. A decrease in levels of heavy oxygen (^{18}O) in cores taken from the ocean floor indicates an interglacial, or warm period.

Research into deep sea cores and ice cores covering the past 1.2 million years or so (dating is not yet precise for the oldest parts of the sequence) has produced a series of twenty-three oxygen isotope stages (OIS for short; also called MIS for marine isotope stage, see Timeline, p. 6), in which the odd numbers represent warm periods and the even numbers represent glaciations. We are currently living in the interglacial OIS 1.

What do these cycles predict for the future of humanity? The Earth's orbit is now in the part of the cycle where its shape approaches a perfect circle, and we can expect the current warm conditions – which are already unusually long and stable – to continue. For example, the 'wobble' cycle is now 6,000 years into a cooling phase and the tilt of the Earth's axis is moving towards a minimum value, which has a cooling effect, but these cycles have been overshadowed by the warmth brought on by the shape of the orbit. Some computer models predict that the next glacial period is at least 5,000 years away.

The Holocene is already the longest stable warm spell since *Homo erectus* walked out of Africa. The effects of human-induced warming on such a complex system are not fully understood, but some climate scientists suggest that rather than being just 5,000 years away from the next cooling phase – which after all is just about the amount of time since the Bronze Age began or since the early phases of Stonehenge in Britain – we may have broken the cycle of glaciations for at least the next 45,000 years.

Cannibals and caves

Some of the most sensational discoveries from Palaeolithic Europe (i.e. the 'Old Stone Age', before the Holocene) have come from the area of Atapuerca in northern Spain since the 1990s. This has marked a change from the previous 150 years, when such countries as Germany, France and Belgium were the focus of attention. One of the directors of the Atapuerca project, Juan Luis Arsuaga, is justified in boasting, 'Today the Iberian Peninsula occupies a very special place in European prehistory.' For starters, it is home to Europe's oldest known human occupation.

The Atapuerca Hills contain an extensive network of caves, parts of which were used as shelters over hundreds of thousands of years. Some of these caves were exposed by an old railway cutting more than 100 years ago by a mining company. Three of the sites in Atapuerca have already altered our understanding of Palaeolithic

Europe: Sima del Elefante is the oldest securely dated site in Europe; Gran Dolina is the second-oldest known site in Europe and has the earliest evidence of cannibalism among hominins; overshadowing even these two extraordinary sites is Sima de los Huesos ('Pit of the Bones'), which has produced the richest collection of hominin fossils anywhere in the world, and which we will examine in the next chapter.

Why is it that we have this extraordinary accumulation of uniquely early sites and fossils in this particular formation in northern Spain? Plant and animal remains from the Atapuerca caves show that the area attracted a wide range of species. The humans shared the area with two-horned rhinos, hippos, bison, sabre-toothed cats, lynxes, bears and hyenas. This diversity probably reflects the different habitats surrounding the Atapuerca Hills, such as river valleys, grasslands and forests. The caves provided commanding views over herds of animals grazing by the confluence of two rivers and thus two migration routes.

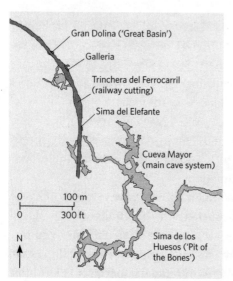

Plan of the site of Atapuerca, in northern Spain. This extensive network of caves was exposed by an old railway cutting, and various sites within the system have yielded extraordinary evidence of early humans, including the oldest known so far in Europe.

40

But such combinations of advantageous topography, habitat diversity and access to resources can be found in many other locations in Europe. The unique archaeological richness of the Atapuerca caves has to be attributed to good fortune in the preservation and exposure of these sites. The sediments did not suffer much destruction from natural processes, such as erosion or tectonic activity. When the roofs of the caves collapsed, the sediments were sealed, and therefore protected, until the railway was cut through them, making long sequences of archaeological layers accessible for excavation. In the case of Gran Dolina, for example, the railway cutting went to a depth of 18 metres (60 ft) of archaeological deposits, exposing layers dating from 300,000 down to nearly 1 million years ago.

In the summer of 2007, the excavations of Sima del Elefante produced a combination of artifacts that is an archaeologist's dream: chipped stone that was clearly crafted into tools by humans (as opposed to stones that may have been chipped by natural breakage); animal bones with signs that they had been cut apart by humans using those stone tools; confirmation of an exceptionally early date (between 1.1 and 1.2 million years ago) from different scientific dating methods; and, perhaps most importantly, the bones of the humans themselves. Many sites have just one or two of these kinds of evidence, leaving archaeologists to argue over key aspects such as dates, the species of hominin present or even whether the site has any conclusive signs of human activity.

The human fossils are a fragment of a lower jaw and a lower tooth. The humans probably picked up large pebbles, found within a few kilometres of the site, and knapped them in the cave, using hard stone hammers to produce simple sharp flakes which they used for defleshing animal carcasses. Some of the animal bones found at the site show cut marks from such tools, as well as percussion marks made in the process of opening up the bones to get at the marrow. The fauna and microfauna suggest that the site was used at a time when the climate was warm and humid, fitting in with the general pattern that humans at this early date had not yet shown a consistent mastery of fire and could not cope with seasonal cold.

The evidence from Gran Dolina is even more abundant than at Sima del Elefante and offers a compelling insight into the lives and minds of these early Europeans. In Gran Dolina, a layer dating to about 960,000 years ago has yielded animal bones, stone tools and fossil human bones belonging to at least six adults and children. Pollen analysis shows that the site was used at a time when climate conditions were wet and temperate. As at Sima del Elefante, the stone tools found in association with the fossils were made of local material and are simple small flakes, some of which were modified and improved, for example by making a serrated edge.

The most intriguing aspect of the Gran Dolina finds relates to the question of how the human remains came to be intermixed with food debris. The human bones are very fragmentary, and most, regardless of the age of the individual, have cut marks made by stone tools in the process of removing the flesh from the bone. These were found along with bones of large, plant-eating mammals, and bones from both groups appear to have been defleshed in a similar manner and to have been cast aside. In other words, Gran Dolina seems to have been a food-processing site, and the humans present were eaten by other humans. How this came to be is open to speculation. A brief look at the practice of cannibalism in our own species points to some likely scenarios.

Cannibalism can be an emotive issue, because in our prosperous modern society it is almost exclusively the domain of deviants, psychopaths or people under extreme duress. Some view the notion that our ancestors practised cannibalism as defamatory. For example, Erik Trinkaus and Pat Shipman in *The Neandertals* (1992) trace how researchers who wished to distance Neanderthals from modern humanity have often accused them of cannibalism. But cannibalism should not surprise us, either in the Neanderthals or in more ancient hominins, because cannibalism is a well-documented part of our own species' behaviour.

Perhaps the most famous modern example of human cannibalism was dramatized in the movie *Alive* (1993). In 1972 a plane crash stranded members of a Uruguayan rugby team and their friends high in the Andes Mountains. In desperation, the survivors resorted to cannibalism of the crash victims as they lived for more than two months far from any source of food until their incredible rescue. But it is not just in acute crises like this that modern humans have been known to eat each other. The behaviour also occurs where there is chronic protein shortage.

As we look back to the distant, pre-agricultural past, access to meat would have depended on the season, success in the hunt, location of wild herds and other factors beyond human control.

Possibly cannibalized human remains from Gran Dolina, Atapuerca, assigned by the excavators to a new species, *Homo antecessor*.

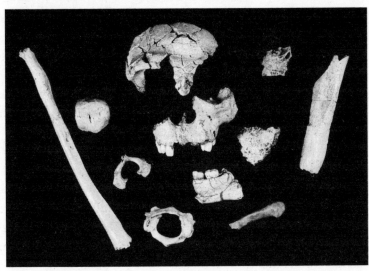

A parallel can be drawn with the South Fore people of Papua New Guinea, as documented in a classic anthropological study from the 1960s by Shirley Lindenbaum and published as *Kuru Sorcery* (1978). The South Fore area is very remote, and it is where both traditional subsistence practices and cannibalism lasted longer than elsewhere in the region. When someone died from causes other than an infectious disease, practically the entire body would be consumed after dismemberment using bamboo and stone tools. This took place in highly ritualized ways designed to honour the dead and their families, with particular body parts reserved for close relatives. The practice only came to an end when the South Fore people suffered an epidemic of a prion disease similar to Creutzfeldt–Jakob disease (the human version of 'mad cow' disease), contracted from the consumption of human brains.

We cannot say whether cannibalism at Gran Dolina took place after an attack by another group, whether it reflected a simple sustenance need or if it was a routine way of honouring those who died in the course of their difficult lives. What we can say is that the act of cannibalism makes the inhabitants of Gran Dolina more, rather than less, human. Also, this site indicates that disposal of the dead by cannibalism probably predates intentional burials and burial rituals by hundreds of thousands of years. In Chapter 5 we will revisit this issue in relation to cases of apparent Neanderthal cannibalism and discuss the continuing controversy.

The first Britons

The site of Pakefield on the east coast of England has been known for its fossils since before the naming of the Neanderthals in 1863. But it was only in the first decade of the 21st century that Pakefield and nearby Happisburgh brought fame to England's East Anglia region for their prominent place in the story of the first human settlement of Europe.

The evidence from these two sites indicates that by around 100,000 years after the cannibalism seen at Gran Dolina, these first Europeans travelled further north than any member of the genus *Homo* had ever gone. The evidence is sparse – with dozens, rather than hundreds, of crude stone tools – but this is enough to prove that hominins had reached a latitude higher than Mongolia and Calgary, Canada.

At this remote time – which the excavators put at 700,000 years ago for Pakefield and between 800,000 and 1 million years ago for Happisburgh – the English Channel had not yet formed, and Britain was connected to the rest of Europe by a land bridge. The area between Pakefield and Happisburgh is the only stretch of England's eastern coastline from this period that is not currently submerged. Located at the bottom of a beachside cliff that is rapidly eroding into the sea, it is lucky that Pakefield has survived at all. When archaeologists unearthed a few dozen stone tools there in 2005, they had to keep a careful eye on the tides to stay safe. It is a reminder of how much important information about archaic humans has already been swallowed by the world's oceans.

Some of the Pakefield tools, which were simple flakes made from river cobbles, were retouched (reworked by being sharpened along the cutting edge). The cores (leftover pieces from which the flakes were taken) were also found. These tools were essentially the same kind of simple cutting implements as those from Atapuerca. They were much easier to make than the handaxes that had long been in use outside Europe.

Pakefield was probably occupied during a mild interglacial. It is clear from the remains of particular rodents and plants that it was used at a time when the summers were warmer and drier than at present and the winters were wet with no frost. As we learned when studying in nearby Cambridge, the people there still would

A flint tool found at Pakefield in Suffolk, England. Tools like this provide some of the earliest evidence for human presence this far north.

have had to contend with very limited daylight hours between November and February.

There are not many large mammals left in Europe, and we now associate such wildlife with nature preserves in Africa. In this period, however, despite its high northern latitudes, England was home to an array of impressive beasts. Pakefield would originally have been in the marshy floodplain of a river estuary, not far from woodland and grassland. This was an ideal feeding ground for hippos, rhinos, bison, mammoths, elephants and deer – all species identified from the animal bones at the site. They, in turn, attracted predatory animals, notably humans, who had an additional incentive to come here – the availability of flint pebbles to make stone tools.

The most startling aspect of the Pakefield stone tools when they were discovered was the unexpectedly early date of 700,000 years ago. This is now fairly secure thanks to a dating device delightfully known as the 'vole clock'. The form of the teeth of various vole species evolved at a pace that turns out to be convenient for dating archaeological layers. The tools at Pakefield are associated with *Mimomys savini*, a species of vole that was also present at Gran Dolina, along with *Mimomys pusillus*, another vole species which seems to have gone extinct about 650,000 years ago. We know, therefore, that humans made these tools slightly before that minimum date.

Before the discovery of Pakefield, the conventional wisdom was that humans at this time were still basically a tropical or sub-tropical species which could not survive so far north. We now know that they were capable of surviving at high latitudes, and perhaps not only when the climate was extremely warm.

In 2005 Happisburgh also began to challenge what we thought we knew about early humans. Located on the banks of the ancient version of the River Thames, Happisburgh produced some eighty stone tools over five excavation seasons until 2010. As at Pakefield, the Happisburgh tools are associated with the extinct voles. But other dating evidence places it even older. The tools came from a layer in which the sand was laid down when the Earth's magnetic poles were reversed. We know from other sites that this occurred 780,000 years ago, so this provides us with a minimum date for Happisburgh. At Gran Dolina, the human remains also come from a layer of reversed polarity.

The pollen traces discovered in the layer indicate that the area was forested and that the climate was cooler than the tropical conditions enjoyed by all humans up to that point. The excavators believe it was sufficiently cool at Happisburgh that the people who made the tools would have needed clothing and possibly also fire in order to live through the winter. Does this mean that the first Britons were also the first humans to wear clothes? This is probably speculating too far, especially because some archaeologists argue that Happisburgh may not be quite as old as the excavators believe. But the lesson of Happisburgh and Pakefield is that our understanding of humans in these early periods is changing fast.

Who were the first Europeans?

The surest way to identify the humans who first colonized Europe is by looking at their bones, the only parts of them that have survived. But in this case, as in virtually every episode in the Neanderthal story, the fossil evidence is sufficiently ambiguous to put consensus on their meaning beyond reach. Here we have two questions. Were the first Europeans an offshoot of Asian *Homo erectus* or were they part of a later wave of migrations from Africa? And are they the ancestors of the Neanderthals?

The earliest human fossils in Europe are from just the sites that we have mentioned: Sima del Elefante and Gran Dolina in Spain. All together, we have fragments of around a dozen individuals, including men, women and children. At Happisburgh, a few dozen footprints were discovered in 2013, but these were not very helpful in identifying the species. We have no human bones from the British sites from this period, so it is simply a reasonable guess that the people there were part of whatever species was in the south of Europe.

Compared with *Homo erectus*, the first Europeans were more modern in a number of ways. Their brains were slightly larger, they were less robust and their teeth were smaller. These traits tend to link the first Europeans more closely to the Africans who were on the evolutionary path to modern humans, rather than to the Asian *Homo erectus* family. To confuse matters, the fossils from Atapuerca, especially from Sima del Elefante, share some of their distinguishing features and bone structure with Chinese fossils from around the same period.

Of the various European fossils from this period, the most surprising is that of a child from Gran Dolina. Its brain is larger than that of an adult *Homo erectus* and its cheekbones are prominent, unlike the flat faces of *Homo erectus*. The excavators said that this child looks very modern and argued that the first Europeans are in the ancestral line of *Homo sapiens*. Given the ancient date of the fossils, the excavators concluded that the people from Gran Dolina were the last common ancestor between Neanderthals and modern humans.

1. Reconstruction of a Neanderthal child based on the Devil's Tower Gibraltar 2 remains excavated in 1926 by Dorothy Garrod, who also made important discoveries in northern Israel and Iraq. Reconstruction by Elisabeth Daynès.

Previous pages
2. Gorham's Cave: this Neanderthal site is just across the sea from Morocco, where early modern humans were living as early as 315,000 years ago. This and other nearby sites have produced some of the latest dates for Neanderthals, although the dates are controversial.

Left
3. 'Excalibur', a handaxe found in the Sima de los Huesos, Atapuerca, Spain, knapped from reddish quartz, which is scarce in this area. This was the only artifact found in a deposit of some 6,500 human bones, representing around thirty individuals, and the excavators believe it may have had symbolic meaning.

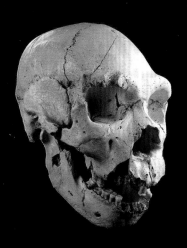

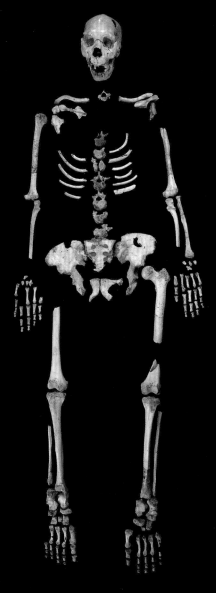

Above and Right
4. Skull 5 from the Sima de los Huesos is the best-preserved early human skull from anywhere in the world; the skeleton (*right*) is reconstructed from the bones of different adults from the Sima. The species represented has been identified by some as *Homo heidelbergensis*, and these remains show incipient Neanderthal traits.

Opposite below
5. A juxtaposition of a Neanderthal skull from La Ferrassie, France (*left*), and a modern human skull from Cro-Magnon, France (*right*).

Following pages
6. Excavations taking place at Amud Cave, Israel, where remains of some sixteen Neanderthals have been found, dating to 50,000 to 60,000 years ago.

7. The cliffs at La Cotte de St Brelade, Jersey, at the bottom of which piles of mammoth and rhino bones were found. Excavators in the 1970s argued that Neanderthals deliberately drove the animals over the cliff in two discrete episodes, which would have required a high level of communication, planning and cooperation.

It is always difficult, however, to compare juveniles of one human species to adults of another. Just as domesticated dogs retain the playful traits of wolf pups in adulthood, one argument goes, modern humans are in a sense domesticated versions of our fiercer ancestors. Could it be that we modern adult humans resemble the pre-adolescent versions of our ancestors? The population of cannibalized humans at Gran Dolina seems to indicate this.

So who were these European pioneers? From the time after *Homo erectus* and possibly also *Homo habilis* left Africa, there are at least a dozen named species whose interrelationships are far from clear. Outside the world of Tolkienesque fantasy literature, we tend to think that it is normal for there to be just one human species on Earth at a time. The past thirty or forty millennia, however, have been the exception, in that we have mostly been the only large-brained primate. Before then, the human world was much more diverse.

Many of the early human species are known from just one individual or from one site, and it is not clear how many true species there were – whether a dozen is an overestimate or an underestimate. It is a classic debate between 'lumpers', who tend to see a lot of variation taking place within interbreeding populations, or who see this variation as reflecting normal differences caused by age, sex and disease within a species, and 'splitters', who are more willing to accept that human diversity involved divisions into several different species.

Given these uncertainties, the excavators of Atapuerca followed the only path left to them, which is to name a new species: *Homo antecessor*. Coined in 1997, this translates as 'ancestor' or 'pioneer' from the Spanish. To confuse matters, there is another possibility, which is that the first Europeans may have been the same species as one identified in Algeria from the same period, known from three jaws and some teeth and assigned to *Homo mauritanicus* in

Excavations at Happisburgh, Suffolk, England, just up the coast from Pakefield. This site, which has yielded around eighty stone tools, has been dated to at least 780,000 years ago.

1954. But the evidence for *Homo mauritanicus* is too fragmentary to make a firm connection with the European material.

On the question of the origins of *Homo antecessor*, the modern features of the child from Gran Dolina are evidence that *antecessor* inherited traits that evolved in Africa after the first human expansion into Asia. But others do see parallels with Asian populations, so this issue seems far from settled.

Equally contentious is the excavators' original idea that *Homo antecessor* is a common ancestor of both *Homo sapiens* and *Homo neanderthalensis*. Arguments against this notion include *antecessor*'s relative 'modernity' in comparison to the Neanderthals. And, to judge from the small number of sites from this period, *antecessor* seems not to have been very successful and probably did not survive Europe's first major glaciation. Recent genetic evidence, however, points to the period 700,000–900,000 years ago as the likely timeframe of the split between the two populations. *Homo antecessor* was in the right place and the right time to be a candidate for common ancestor.

Antecessor may have been at least closely related to the population that gave rise both to Neanderthals and modern humans. In the next chapter, we will keep the focus on Atapuerca, where we will meet a population with a more certain claim to being the Neanderthals' direct ancestors.

Defeating the Cold

600,000 to 250,000 years ago

Starting just over 650,000 years ago, Europe underwent almost 40,000 years of icy conditions in the coldest and most sustained glacial period since humans first arrived. For a genus that was adapted to the tropics, this probably marked the end of what was never more than a low-level colonization of the continent.

After the melt, a new kind of human, bearing such tools as handaxes and wooden spears, appeared in Europe. These humans fended off other large carnivores to obtain primary access to large game, such as the rhinos and horses that they were probably hunting, and seem to have shown respect for the dead. Their occupation was much more extensive than that of *Homo antecessor* before them. The next time the climate deteriorated, they were able to maintain a presence in Europe, living through increasingly long and hostile glaciations. This second wave of Europeans evolved into a new form of humanity which, by 250,000 years ago, had become recognizably Neanderthal.

Who were these more successful European colonists who gave rise to the Neanderthals? Three skulls from Ethiopia, Zambia and South Africa provide a clue. Dating to the period covered in this chapter, these skulls show similarities to the species that made its first appearance in Europe just over half a million years ago and probably shared a recent common ancestor. Some classify this species in Africa as *Homo rhodesiensis, Homo bodoensis* or 'archaic'

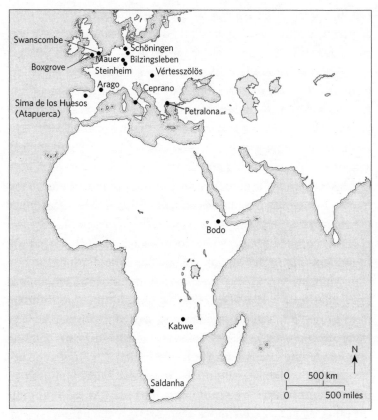

Map showing the major sites discussed in this chapter and attributed to *Homo heidelbergensis* in Europe and *Homo rhodesiensis* or *Homo bodoensis* in Africa.

Homo sapiens. For a time they were seen as the same species as their European counterparts, and because priority goes to the first named fossil of its kind, both the African and European fossils were called *Homo heidelbergensis.* Now, as the African situation seems more diverse than previously thought, for most researchers *Homo heidelbergensis* just refers to the European variety. The name of their common ancestor is, for now, unresolved but may be something like the *Homo antecessor* we met in Chapter 2. (Even in Europe, there is doubt whether *heidelbergensis* has the coherence of a separate species and it may soon just be called proto-Neanderthal. We will continue to use this name which has been in use for several decades.)

With two prominent human species named after places in Germany, it is easy to forget that Africa was the primary engine of human evolution. Neanderthals, as we saw in Chapter 1, are named after the Neander Valley near Düsseldorf. Down a road called the E31, less than 300 kilometres (185 miles) away, is Heidelberg, an ancient seat of European learning and, more recently, home to American and NATO military headquarters in Europe. Heidelberg received the honour of having a human species named after it when a jawbone turned up in a quarry in the area of nearby Mauer in 1907.

Few people have heard of *Homo heidelbergensis*, despite its prominent role in the human story. If the Neanderthals are our closest human cousins, then *Homo heidelbergensis* is something akin to a favourite uncle. Fossil evidence suggests that this species arose from a population that had a presence across Africa and Europe more than 600,000 years ago. Some researchers believe it reached into Asia, citing fossils from China and India.

Looked at globally, it is possible that *Homo heidelbergensis* was part of a second major wave of migration out of Africa, following the initial *Homo erectus* exodus. By the time of the final wave – of our own *Homo sapiens* ancestors – the European *heidelbergensis* population had evolved into Neanderthals. In this scenario, the initial *Homo antecessor* colonization of Europe probably represents a minor, less successful migration from Africa in the time between the *Homo erectus* and *Homo heidelbergensis* expansions. Another possibility is that *Homo antecessor* evolved locally into *Homo heidelbergensis* in Eurasia, and the three African skulls seem similar but were evolving on a separate path to *Homo sapiens*. Small groups were probably going in and out of Africa throughout much of the last two million years, and it was only a few times that these events coincided with evolutionary changes and left evidence that we can read in the fossil record.

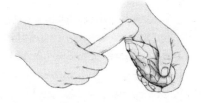

Making a handaxe with a bone or antler tool.

The *Homo heidelbergensis* occupation of Europe is framed by two major changes in stone tool technology. They used handaxes, in contrast to the simple pebble-and-flake tools of *Homo antecessor*. By about 250,000 years ago, which most researchers agree marks the appearance of Neanderthals, there is another change, as cutting tools became smaller and more sophisticated. As we will see in the next chapter, this later technological shift seems to occur simultaneously in Europe, western Asia and South Africa.

In contrast to the earlier *Homo antecessor* occupation of Europe, the evidence of *Homo heidelbergensis* is widespread, and we know a great deal more about *heidelbergensis*'s evolution, anatomy and behaviour. As with our understanding of *Homo antecessor*, our knowledge of *Homo heidelbergensis* is exploding thanks to a wealth of new discoveries. The last few decades have seen the excavation of the richest hominin site in the world at Atapuerca, the discovery of the world's oldest wooden artifacts in the form of spears from Schöningen, Germany, and the excavation of the best-preserved butchery site from this period at Boxgrove, England.

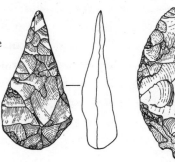

A pointed handaxe from Swanscombe and an ovate handaxe from Boxgrove, both sites in England (lengths approx. 140 mm and 160 mm).

As we gain a fuller understanding of the Neanderthals' immediate ancestor, it is remarkable how familiar it is in behaviour and appearance. In Europe, half a million years ago, there was an intelligent, capable and even sentimental form of humanity. *Homo heidelbergensis* is closely related to us too, and its essential modernity should not surprise us. The more interesting question is how Neanderthals evolved from *Homo heidelbergensis*, to what extent they followed a parallel course to our own direct line and in what ways they differed. In this chapter we examine our shared starting point, remembering that the evidence from this time is far richer in Europe than in Africa, and so it is here that we find the clearest understanding of our own ancestry as well.

A Palaeolithic Pompeii

In a quarry at Boxgrove in southern England, archaeologists excavated a 500,000-year-old ground surface that was so well preserved that they could make out the exact spot where an ancient human sat to knap a handaxe. Starting in 1985, a decade of excavations produced no fewer than 300 handaxes, along with butchered remains of large animals such as elephants, rhinoceroses, horses, bison and red deer.

How did the remains of butchery and flint-knapping come to be so well preserved at the site? Boxgrove is located beneath a cliff that once marked Britain's southern shore. Today, it is quite a way inland. When the sea level began to fall as a result of growing ice sheets, the area became a salt-water lagoon and the emerging coastal plain proved attractive to animal herds. Meanwhile, the cliff itself was a good source of flint to make handaxes. With large game and raw material to make tools in a confined area, Boxgrove seemed custom-made for human exploitation.

The lagoon was the key to the site's preservation. In a similar, albeit much less dramatic, way to how life in Pompeii was frozen in time by the ash of a volcanic eruption in AD 79, there are a number of flint-knapping and butchery episodes at Boxgrove around 500,000 years ago that were preserved by the accumulating silts of the lagoon. Thanks to these shallow waters, we are able to see the results of activities that were probably as brief as fifteen minutes and no longer than an hour or so. To be able to see such a snapshot of ancient life is incredibly rare, but at Boxgrove it is also unfathomably distant in the past.

The great excitement from Boxgrove, however, was not just the preservation but the fact that the site has overturned some long-held ideas about the capabilities of our recent ancestors. It is clear from the animal bones that humans were the first predators to reach them. By the time scavengers such as hyenas and wolves fed on the bones and left their teeth marks, there were handaxe cut marks on them already. And these bones formed complete animals, meaning that the carcasses were still intact when humans started defleshing them.

In other words, whether or not humans hunted and killed such enormous game, they were at least able to drive away rival predators and have their choice of meat and fat. They may well have directed the animals into the lagoon to trap and kill them. Until recently, some researchers thought that the Neanderthals were merely marginal scavengers, making do with the scraps of other animals' kills, so it has come as a surprise that even the Neanderthals' predecessors were probably hunting large animals such as elephants and rhinos and were able to drive away lions and wolves, whose bones have also been found at the site.

To put an end to the question of human scavenging versus hunting at this early date, researchers found a 'smoking gun' in the form of a horse shoulder blade that showed signs of being pierced by a spear. In 1997 startling confirmation came when actual examples of such spears, some still embedded in horses, were uncovered from the site of Schöningen in Germany and dated to around

340,000 years ago. There is little doubt that *Homo heidelbergensis* was a formidable hunter. At the end of this chapter, we discuss the European *heidelbergensis*'s practice of short-range ambush hunting and consider whether this may have been a pivotal factor in the evolutionary split between the Neanderthals and modern humans.

Towards the end of the Boxgrove project, a new discovery rattled the archaeological world once again. In May 1994 *Nature* published news that a shin bone found at Boxgrove was the same age as the jawbone from the *heidelbergensis* 'type-site' of Mauer, making it one of the oldest fossils then known in Europe. This claim to fame was not long-lived, because bones almost twice as old were discovered that very summer at Gran Dolina (which we discussed in Chapter 2). Known as Roger after discoverer Roger Pedersen, the original owner of this shin bone is still on record as the most ancient Briton. Two teeth turned up at Boxgrove a year later, confirming that the Boxgrove people were *Homo heidelbergensis*.

From these few fossil remains we can get a different kind of snapshot of life at Boxgrove. We knew from the preservation what the humans were doing at the site. Now we know much about who some of them were. The owner of the shin bone was built like an athlete. The individual was probably a male and lived to the ripe age of forty. Estimates of his height and weight (180 cm and 90 kg, or 5 ft 11 in. and 200 lb) are eerily close to those

Reconstruction of a male, nicknamed Roger, from the site of Boxgrove, based on the robust shin bone excavated at the site as well as other fossils of *Homo heidelbergensis*. He is equipped with a wooden spear similar to an actual one discovered at Schöningen in Germany.

of the heavyweight champion boxer Mike Tyson. As for the teeth, the cut marks indicate that they belonged to a right-handed person who used his jaw as a vice for cutting.

The animal species at Boxgrove place the site in the warm period just before a cold spell called the Anglian glaciation, when most of northern Europe was covered by ice sheets or was a polar desert. Humans did not survive in Britain in these conditions, but they did survive on the continent.

The world's first mass grave?

Within a dark and forbidding cave at Atapuerca, there is a seemingly bottomless hole. Some 14 metres (46 ft) down a narrow shaft lies a small chamber. First explored by palaeontologists in the 1970s, the chamber started to produce the world's greatest collection of ancient human fossils – both in terms of quantity and preservation – in the 1990s. Appropriately called Sima de los Huesos ('Pit of the Bones'), this site represents a dual mystery: who are the unfortunate souls that came to rest in the Sima, and how did they get there?

At first this site was dated to 250,000 years ago, but this has been slowly pushed back through successive advances in dating technology. More recently, the excavators put it at 430,000 years ago. The bodies in the Sima already show incipient Neanderthal features, making them among the earliest that we can attribute with any confidence to the ancestral line of the Neanderthals. Some researchers, such as Chris Stringer, go so far as to say that they are early Neanderthals, a point of view that gained momentum in 2016 when a team from the Max Planck Institute for Evolutionary Anthropology in Leipzig, Germany, led by Matthias Meyer, managed to extract nuclear DNA from bones at the site. Given the age of the bones and the fact that their morphology was not fully Neanderthal, other researchers, such as Juan Luis Arsuaga, the Spanish excavator, regard them as late *Homo heidelbergensis*. Because proponents of both viewpoints agree that *heidelbergensis* evolved into *neander-thalensis*, this can be seen as something of a semantic point.

Excavation at the bottom of the cramped pit has been pain-staking but rewarding. The site has produced over 6,500 hominin fossil bones and teeth, representing around thirty individuals, plus fossil animal bones of hundreds of cave bears and other predators such as lions, wolves and foxes. How did so many humans end up in such a small place? In contrast to nearby Gran Dolina, there are no remains of animals that might represent food debris. This was not a simple dumping ground.

There is a clue to this mystery in the only artifact found among the bones. It is a beautiful handaxe, knapped from reddish quartz, a rare material among the river pebbles used for making stone tools in the Atapuerca area (see plate 3). The excavators have named this handaxe Excalibur, in order to drive home their interpretation that this was an artifact loaded with symbolic meaning and that what took place at Sima de los Huesos was some sort of funerary rite. These are extraordinary claims for such ancient hominins, who until recently were thought not to have had the capacity for symbolic thought or to have attached any special meaning to death. But Sima de los Huesos is an extraordinary site, and it may just warrant an extraordinary interpretation.

Another clue is the age profile of the dead. Thanks mainly to the amazing preservation of the teeth, we know that most of the individuals died in their prime. If the Sima were something like a community grave site, then we would expect the bones to be mainly of children and the elderly, who typically die in greater numbers. But of the thirty or so individuals buried there, we have only one pre-pubescent child and just three adults over thirty years old.

The excavators believe that a catastrophic event, such as a drought, drove these hominins to the refuge of the Atapuerca Hills. In this scenario, only the fittest members of the group made it as far as the cave. Then, when their luck ran out, the last survivors dragged their deceased friends and family deep into the cave and dropped them in the pit, along with a symbolic red handaxe.

This is an incredibly moving hypothetical story that spotlights the essential humanity of these ancestors of the Neanderthals. A close look at the fossils reveals how much like us they were. For example, in contrast to earlier hominins who were more balanced in their handedness, they were mostly right-handed, which we can see from their stronger right arm bones and leg bones, and also from the imprint of their brain shape on the inside of their skulls. The population at the Sima also seems to have been keen users of toothpicks, which have left grooves on their teeth.

The difference in body size between men and women (sexual dimorphism) at the Sima is similar to that in our species. This is a sign of modernity because in non-human primates and early hominins, males are much larger than females, while in modern humans the males are only slightly larger. Many researchers believe that the reduction in sexual dimorphism accompanied a shift from a male-dominated harem-type mating arrangement to a simpler, two-females-for-one-male polygamy or even pair-bonding monogamy, as predominates today. In short, the people of the Sima were likely right-handed individuals who entered into arrangements similar to our nuclear family units and placed an emphasis on dental hygiene.

At the same time, the men and women of the Sima did not look or behave quite like us. Their thick leg bones show that they were very muscular, probably even more so than the classic Neanderthals of 60,000 years ago, who had been regarded as the most heavily built human ancestors. The people of the Sima had bodies like the strongest modern rugby players or American football players. Many of them had brains that were as large as ours, but flatter. Their skulls were long, with low foreheads, prominent brow ridges, powerful jaws and no jutting chins. From wear patterns on their teeth, they must have used their jaws as a 'third hand' to hold meat, skins or plant material while they used their free hands for cutting.

At the Sima we have an early example of the tension between the physical uniqueness of the Neanderthal form and their behavioural similarity to early *Homo sapiens*. According to Antonio Rosas (see plate 16), a palaeoanthropologist who launched his career studying the human jawbones at the Sima, 'Neanderthals are, let's say, "a distinct biological entity" (distinct from modern humans), whose anatomy and, apparently, also physiology are visibly different from ours. But, on the other hand, the more data we gather on their behaviour and reproductive biology, the more similar they seem to be to what we call the "modern human pattern" (something perhaps weakly defined). What is behind these two extremes of reality is something that, for me, is becoming challenging to reconcile.' Resolving this tension is central to understanding what it means to be human.

One definition of humanity is our capacity to care for the old and the sick. For this reason, perhaps the most noteworthy individual in the ossuary is represented by Skull 5, an elderly person (which, in Palaeolithic terms, means over fifty years old) who had serious dental problems and a brain size smaller than the modern human range. This is the best-preserved fossil human skull from anywhere in the world (see plate 4). He (for we presume he was a male, judging by the proportions of the face) must have had trouble chewing his own food. His presence in the Sima is just another part of the mystery of how this population came to be preserved here.

It is impossible to tell from the stratigraphy whether the bones accumulated as a result of a single catastrophic event, as the excavators suggest, or more gradually. One can think of all sorts of scenarios in addition to the idea of an ecological crisis. Perhaps the single child in the group wandered into an uninhabitable cave and fell in a pit. Then one by one the rest of the clan became trapped in the course of failed rescue attempts. Or perhaps the young adults were adventuresome cave explorers who died in a series of tragic events over many generations.

Other researchers look to a more conventional alternative – that the bodies, along with cave bears and other predators, were first deposited elsewhere in the cave system and then washed down into the Sima as a mixed, disarticulated mass. The archaeologist Paul Pettitt describes this as an early example of funerary caching, or the abandonment of human corpses in a preferred location. This theory, however, does not quite explain the relative absence of young and old or the fact that the skeletons seem to be complete. Sima de los Huesos, one of the earliest known sites of the Neanderthals' ancestors, is an ancient mystery that still holds many secrets.

THE PETRALONA AFFAIR

Locked in a vault at the University of Thessaloniki in northern Greece, shielded from the controversies that surround it, lies one of the best-preserved *Homo heidelbergensis* skulls ever found. When locals on the nearby Halkidiki Peninsula uncovered it in Petralona Cave in 1960, it caused a stir. But few could have foreseen how this key piece of the puzzle of Neanderthal origins would be the focus of a decades-long legal dispute.

The protagonist of this tale is Aris Poulianos, whose career in Greece contains no academic appointments outside of the Anthropological Association of Greece, an organization he founded and presided over for decades until he handed it to his son, Nikos. From this base, he came to have complete control over the excavation rights as well as tourist access to one of Europe's most important Palaeolithic sites. To understand how this came to be, we must delve into the polarized political history of a country that still bears the scars of the Second World War and a subsequent civil war between rival resistance factions.

According to the Anthropological Association's website, Poulianos fought with the communist resistance to Nazi occupation. After the war, he attained a PhD on 'The Origins of the Greeks' in Moscow before he returned to Greece in 1965 in the hope of studying the Petralona skull. He soon entered the civil service as a scientific adviser. In 1968, using his position as vice-president of the Greek Speleological Association, he began digging in Petralona. The Archaeological Service soon put a halt to the project, however, and Poulianos was expelled from the Speleological Association.

The reversal of fortune that handed the site back to Poulianos was a consequence of the reaction to the

Cast of the skull found at Petralona Cave on the Halkidiki Peninsula in Greece.

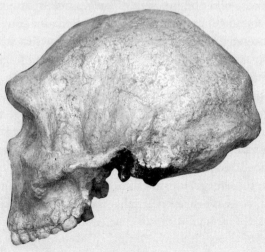

right-wing military junta that had ruled Greece from 1967 to 1974. The junta was notorious for its repression of academic freedom, and Poulianos, who had been imprisoned in the early days of the regime, could portray himself as a scholar oppressed for his political views. In 1974 the newly restored democratic government granted him an excavation permit for Petralona, and in 1979 he signed an agreement with the Greek National Tourism Organization to conduct further excavations and develop the site as a tourist attraction. In 1983, under a left-wing government, the Greek state expelled Poulianos from Petralona a second time, and he turned to the courts. The government alleged that Poulianos was not excavating in a scientific manner and was destroying the site.

Poulianos has sought to portray the skull and the site in superlative terms. First, he insisted that the skull is the oldest in Europe. In a paper in 1971, he dated it to 70,000 years ago, which he claimed made it the oldest known at the time. In 1981 he added a zero to the date, making it 700,000 years old, maintaining its most ancient status in the light of more recent discoveries. Poulianos retrospectively changed the stratigraphy of the find, moving it down from Layer 10 (as he said in 1971) to Layer 11, a level he originally claimed was empty of any human remains or artifacts.

Scientific dating techniques now place the skull between 160,000 and 620,000 years old, and the prevailing view is that the true date lies close to the midpoint between these extremes. When Chris Stringer, of London's Natural History Museum, argued for this date at a conference in 1988, Poulianos rushed the stage and had to be restrained (as recounted in James Shreeve's excellent book *The Neandertal Enigma*).

Poulianos's strong feelings about the skull's age are but one part of an increasingly idiosyncratic view of the prehistoric past. He has used the skull to name a new species, *Archanthropus europeus petralonsiensis*, which he alone recognizes, and which he believes is somehow ancestral to the Sarakatsani, a modern population of nomadic pastoralists who inhabit the same area. He claims to have uncovered bone fragments from fifteen other individuals in Petralona, although he has never published evidence of this. What he has published are pictures of tools associated with the skull, but these do not appear to show evidence of being anything other than unworked rocks.

With the entry of Poulianos's son onto the scene, the Anthropological Association of Greece made more incredible announcements. It claimed to have found evidence in Petralona and the nearby site of Nea Triglia of a sculpted figurine from 500,000 years ago, fire from 1 million years ago and another new species, *Homo trigliensis*, on the basis of 10- to 11-million-year-old stone and bone tools. These dates are at best controversial – the last one outlandish – and have not found support in international peer-reviewed periodicals.

In 1996 Poulianos won his legal battle with the Greek government and, for reasons of preserving 'intellectual freedom', was able to take total control of Petralona the following year. The Ministry of Culture tried on numerous occasions since then to regain possession of the cave, and finally succeeded in 2011. The Petralona skull, meanwhile, remains under lock and key.

Becoming Neanderthal

One can feel spoiled after looking at such sites as Sima de los Huesos and Boxgrove. In contrast, the human fossil record over the next few hundred thousand years, leading up to the appearance of unmistakable Neanderthals, is fragmentary and not well dated. Each of the fossils from this time has a unique mixture of Neanderthal- and *erectus*-like traits, making it difficult to map an evolutionary trajectory from *heidelbergensis* to the Neanderthals.

In the absence of good dates, it is common to estimate ages based on how Neanderthal-like fossils look. As dating methods have improved, it has become clear that Neanderthal traits appeared earlier than previously thought. The bones from this period are frequently revised back in time, sometimes by hundreds of thousands of years. It is likely, as more dates become available, that there are surprises still to come.

One key unresolved question is whether the whole population of Europe evolved into *neanderthalensis*, or whether there was more than one species present. Some researchers seem tempted by the two-species scenario. Chris Stringer has noted a suggestive difference in stone tools, whereby handaxes and flake tools (the latter in an industry called the Clactonian, named after Clacton-on-Sea, a seaside resort in Essex, England, where such tools were found) are never found in the same archaeological layer, and thinks this might indicate two separate populations.

Might the 'Clactonian people' represent a less successful branch of the human family tree, separate from the 'handaxe people', who were evolving into Neanderthals? We remain sceptical that the absence of handaxes – a tool that was used by a number of different forms of human – might indicate a separate species. Yet the pattern of fossils does seem to allow for the possibility that what we attribute to the *heidelbergensis* population of Europe was really two species.

What do the bones tell us about the origins of the Neanderthals? Among the oldest dated fossil remains of *Homo heidelbergensis* are the lower jaw from Mauer that gave the species its name and the shin bone and two teeth from Boxgrove, all about half a million years old and too fragmentary to be very revealing. The cache at Sima de los Huesos may also be very old, as may the skull from Petralona, which was the most complete known *Homo heidelbergensis* skull before Skull 5 was uncovered at the Sima. Like so many *heidelbergensis* fossils, the Petralona skull has its own particular mixture of modern and archaic traits and, as we have seen, its exact age is unknown.

The face of the Petralona skull looks even more Neanderthal-like than Skull 5, while the back of its head is closer to *Homo erectus*. At the front of the skull, its massive brow ridge is reminiscent of *erectus*, but unlike the heavy 'unibrow' of *erectus*, it is individually arched above each eye socket and is hollowed out by large air chambers called frontal sinuses, similar to the Neanderthal form. Its cheekbones and large nasal opening anticipate features of the Neanderthals, but its face is broader and flatter than in Neanderthals. Its cranial capacity (*c.* 1220 ml, 41 oz) sits on the margins between the *erectus* and Neanderthal ranges.

A braincase from Ceprano, Italy, while incomplete, shows a similar mixture of *heidelbergensis* and more archaic traits, and for a time was thought to be older than the date of 400,000 years ago that it now has. Also from around this time are fragmentary fossils from the site of Arago in the French Pyrenees. Some archaeologists believe that the inhabitants of Arago Cave were hunters of wild sheep, whose bones are also found at the site. Some fragments at Arago have been fitted together to reconstruct the face and right side of a skull that appears similar to the Petralona skull and is transitional between an *erectus* and a Neanderthal shape. A hip bone and other body bones confirm the trend from Boxgrove and the Sima, in which *heidelbergensis* has a very robust, yet modern-looking body.

It is worth pausing here to consider why the European *heidelbergensis* population was so large and muscular, with the bodies of boxers. One theory is that they had become specialized ambush hunters, using wooden spears in close combat against sizeable prey. This strategy contrasts with that of our own African ancestors, who had the ability to run down prey over long distances. Contemporary African hunter-gatherers have been observed using this hunting technique, in which hunters effectively out-endure prey, which eventually become tired and bewildered, making a kill relatively easy after a long run. Pursuit hunting can be effective in the hot African savannah, which is mostly flat and free of trees, but even in a partially forested and more temperate European river valley, it would not work. It is possible that this divergence in hunting strategies explains the difference in physique between Neanderthals and modern humans.

Following an extremely cold glaciation and during a long warm period called the Great Interglacial, we find a skull in Britain from 400,000 years ago that resembles the skulls of the Sima. A quarry near Swanscombe, England, yielded three skull fragments which, despite being found in 1935, 1936 and 1955, belonged to the same individual. The Swanscombe skull has a little dimple, called a suprainiac fossa, at its base, and this feature is typical of Neanderthals; it probably has something to do with the way that strong neck muscles attached to the back of the skull. Unfortunately, the face of this skull has not been found, but we do know that the estimated brain size (1300 ml, 44 oz) was larger than that of the Petralona skull.

A skull from Steinheim, Germany, some 100,000 years after Swanscombe, challenges the notion of a smooth evolution from the earliest European *heidelbergensis* fossils to fully formed

Neanderthals. Found in 1933, the Steinheim skull is smaller (1100 ml, 37 oz) than Swanscombe and Petralona, and its mode of 'neander-thalization' is reversed from the pattern at Petralona and the Sima, where Neanderthal traits appeared earlier in the front of the skull. With the caveat that the Steinheim skull is incompletely preserved and its front half is distorted, the back of the skull looks much more Neanderthal-like than the front.

Two sites roughly contemporary with Steinheim – Bilzingsleben, Germany, and Vértesszöllös, Hungary – provide a more profound challenge to the idea of a gradual, uniform evolution towards Neanderthal forms. At Bilzingsleben, remains of three human skulls were unearthed in the 1970s. At Vértesszöllös, a single skull fragment was discovered on 21 August 1965, the traditional 'name day' or religious festival day for the prophet Samuel, and this most ancient Hungarian was given the popular nickname Samu. Ian Tattersall has argued that skull fragments from these two sites do not seem to have any incipient Neanderthal traits at all and are associated with flake-based tools such as the Clactonian. Stringer believes that they are related to *heidelbergensis*, but describes them as more primitive.

Could it be that while part of the European *heidelbergensis* population was evolving into Neanderthals, another part was not? Or perhaps more archaic humans periodically entered Europe from Asia, as the local *heidelbergensis* people established their own evolutionary trajectory. This scenario may be supported by mitochondrial DNA evidence retrieved from human bones at Sima de los Huesos, linking that population with Denisovans in Asia. There is some poignancy in the thought that when fully formed Neanderthals appeared some 250,000 years ago, they had recently out-competed a more archaic European population. If that is the case, then the Neanderthal story is framed by two encounters with other forms of human. In the first encounter, the Neanderthals emerged victorious as the sole occupier of Europe. In the second, they ceded the continent, and the planet, to us.

By 250,000 years ago, as the successful 'handaxe people' took on an increasingly Neanderthal-looking head shape, their stone tool technology became more advanced and they relied less on the bulky handaxe. It is at this moment that the Neanderthal story truly begins.

Meet the Neanderthals

250,000 to 130,000 years ago

With brains, size is not everything, but it does matter. Since the beginning of the genus *Homo*, the human brain had been evolving larger capacities. From around 250,000 years ago, we have a human skull from Reilingen, Germany, that housed a brain with a larger volume than that of perhaps half the people who will read this book. It is this moment, when the modern brain size threshold is reached in Europe, that most researchers mark as the beginning of the time of the Neanderthals.

What does brain size tell us? On its own, it does not tell us very much. But it is part of an intriguing pattern from this period which includes evidence of large-scale cooperative hunting and the manufacture of more complex stone tools. These developments point to a significant advance in the behaviour of the first Neanderthals. This advance is so interwoven into our own daily lives that it can be easy to forget just how revolutionary it was when it appeared. This behaviour, which separated the Neanderthals from their *heidelbergensis* ancestors and distinguished them from almost all other human species, can be summarized in two words: forward planning.

Forward planning requires imagination – the ability to envision what a future situation will be and what intermediate steps are needed to bring that situation into being. It also requires understanding – the knowledge that animals, plants, rocks and other humans all have behaviours or properties that are predictable and consistent. In many cases, forward planning requires the kind of cooperation that can only come about through complex communication.

All hominins must have possessed the ability to plan ahead to at least some degree. By 250,000 years ago we have evidence of a leap forward, an extension in the number of steps ahead that humans could reliably plan and the distances they travelled and carried materials. While we can see this only in evidence of hunting and toolmaking, we can safely assume that it applied to other areas as well.

One significant area where forward planning would have given Neanderthals an advantage is in social relations, enabling them to increase their group size and their range. For anthropologist Robin Dunbar, who developed what he calls the social brain hypothesis, the reproductive advantages derived from successfully navigating complex social relations have been the main driver of the increase in brain size in our evolutionary past. We can see that larger brains could also help in improving food procurement through more efficient hunting.

All of these strands of evidence – brain size, bone morphology, group hunting, stone tool technology – enable us to look back to 250,000 years ago as the moment that marks the full emergence of the Neanderthals. At the same time that the European Neanderthals were evolving a distinct body form, especially evident in the shape of the skull, a rival population in Africa was making similar cognitive leaps forward. The Neanderthals of Europe and the archaic *Homo sapiens* (which some call *Homo rhodesiensis* or *Homo bodoensis*) of Africa were both progressing in parallel, and neither could yet claim any advantage over the other.

Neither of these groups was static, of course. The populations did not become 'Neanderthal' or '*Homo sapiens*' overnight, and we cannot pinpoint a true start date. We have chosen 250,000 years ago,

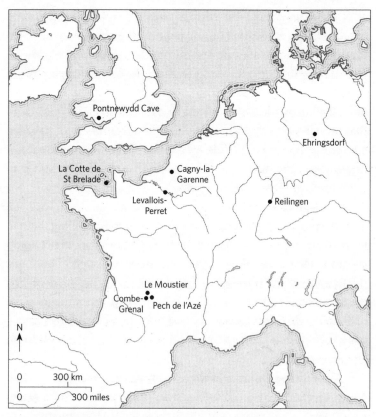

Map showing the key Neanderthal sites discussed in this chapter, which have evidence for some of the earliest Neanderthal behaviour and body shape.

like many researchers, as an important point along a continuum of hundreds of thousands of years in which body types trended towards distinct forms and behaviour became more complex. Neanderthals continued to evolve into what is termed 'classic' or 'late' Neanderthals. We use similar terms to describe the parallel developments in Africa, which is why we favour calling that species archaic *Homo sapiens* (and we will avoid the oxymoronic 'archaic modern humans', though that is essentially what they were).

During the period covered in this chapter, Neanderthals and archaic *Homo sapiens* – both now distinct and more advanced than *Homo heidelbergensis* – were flourishing in their separate homelands. Yet there is little evidence that either species had spread into Asia, and they still had significant changes ahead of them. By the time we reach 130,000 years ago, which we discuss in Chapter 5, there is evidence of Neanderthals and *Homo sapiens* expanding into Asia, where they competed for resources and territory.

The first Neanderthals

The fossil record from 250,000 years ago is sparse, but the few human bones that do survive indicate that a major change had taken place in Europe. This change is most evident in the skull.

In 1978 fragments of an ancient cranium turned up in a gravel pit in Reilingen, Germany. The animal bones found with it suggest a date older than 200,000 years ago. Researchers have measured these fragments in every conceivable way and argued about the species they represent.

There is agreement that certain features – especially a small depression and fold in the bun-shaped back of the skull – place it well on the evolutionary path to Neanderthals. Some, such as Ian Tattersall, describe it as *Homo heidelbergensis* with strong Neanderthal affinities. Others, such as Juan Luis Arsuaga, argue that this skull is among the first 'true and complete Neanderthals'. Because evolution and change are constant features of the European fossil record, this can be viewed as a semantic point. Taking into account the contemporary behavioural developments we outline in this chapter, we think this skull is an appropriate one to consider as among the first of the Neanderthals. In addition, the one measurement that, for us, is the most telling is the estimate of this individual's cranial capacity.

How can you estimate the size of an ancient individual's brain, when you only have part of the skull? Researchers used more complete skulls from later periods to model the likely shape of the missing pieces. Then they created a silicone rubber mould of the inside of the skull and put it in a tub of water to measure displacement. They did this a few different times using different assumptions and came up with an average result of around 1430 ml (48 oz). This is an extremely large brain for such an ancient human skull – the largest one ever found up to that age anywhere in the world. *Homo heidelbergensis* brains range from around 1100 ml up to 1350 ml or so (37–45 oz). The Reilingen brain sits right on the average of modern *Homo sapiens*. Later, Neanderthal brains got even larger.

There is another skull from Germany dated to the same period. Discovered in a limestone deposit near Weimar, the Ehringsdorf fossils – which include various remains of more than one individual in addition to skull fragments – show a mixture of advanced features, such as a high skull and forehead and a large brain, with more primitive ones, such as a strong brow ridge. Discovered in the early 20th century along with animal bones, invertebrates and plant remains that indicated a temperate climate, they were thought to be about 120,000 years old – the last time that the climate in Palaeolithic Europe was as warm as in modern times. This seemed to make sense because everyone agreed that by this time the Neanderthals had been fully formed, and the morphology of the fossils was Neanderthal-like enough that it did not contradict that age estimate. But more recently, dating with newly available scientific methods has pushed the site and the fossils back to about 230,000 years ago, placing them in an earlier warm stage of the Palaeolithic.

Two additional skulls were also recently redated, giving further evidence that the Neanderthal form was established by around

250,000 years ago. They are from Saccopastore, an area just outside Rome, Italy, where they were discovered in a gravel pit in 1929 and 1935. They are not as robust as classic Neanderthals, and their brain capacities were smaller than the Reilingen individual. With Reilingen, Ehringsdorf and Saccopastore we have good evidence that by this time, the people of Europe were well on the way to taking the form of the 'classic' Neanderthals that came later and for which we have quite a lot more fossil evidence. At this early Neanderthal period, what can we say about the distinct Neanderthal form?

The signature Neanderthal characteristic was prominent ridges above each eye. This contrasted with the single brow ridge of *Homo erectus* and the near absence of brow ridges in *Homo sapiens*. These ridges probably made Neanderthals look quite fierce. There is no getting past the fact that to a modern human, Neanderthals probably looked ugly. In contrast to the high cheekbones and prominent chins we find so irresistible to gaze upon, Neanderthals had enormous, broad noses, faces that jutted forward and no chins.

At the back of their heads, they had what are called an occipital torus and a suprainiac fossa, which are technical terms for a little protuberance with a small pit. These features probably came about from the strong muscles needed by Neanderthals, who used their jaws like vices or a third hand. This interpretation is supported by their teeth, which tend to be very worn down. Putting all this together, their heads were large, but flatter than ours. They seemed to thrust forward and back, while our heads are more globular with high foreheads. Recent research has shown that both Neanderthals and modern humans have elongated brains at birth, but modern humans develop globular-shaped brains (and skulls) during the first year of life, while the Neanderthal skulls retained their original shape. One of Dimitra's students offered this analogy, referring to Neanderthals' muscularity in contrast to our gracile build: if modern humans look like football (soccer) players, the Neanderthals were more akin to rugby players. We can push this analogy further: if a modern human skull, with its rounded top and back, resembles a football (soccer ball), a Neanderthal skull, with its elongated shape, shallow forehead and protruding bun at the back, resembles a rugby ball (or, indeed, an American football).

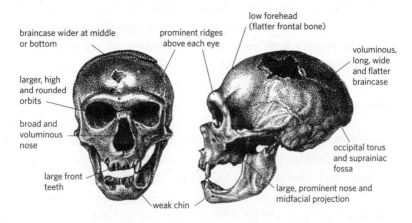

braincase wider at middle or bottom

prominent ridges above each eye

low forehead (flatter frontal bone)

voluminous, long, wide and flatter braincase

larger, high and rounded orbits

broad and voluminous nose

large front teeth

weak chin

occipital torus and suprainiac fossa

large, prominent nose and midfacial projection

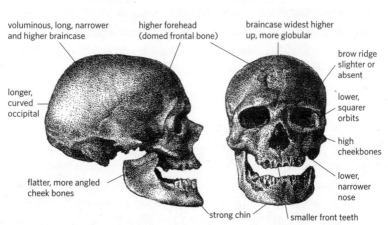

voluminous, long, narrower and higher braincase

higher forehead (domed frontal bone)

braincase widest higher up, more globular

brow ridge slighter or absent

lower, squarer orbits

longer, curved occipital

high cheekbones

lower, narrower nose

flatter, more angled cheek bones

strong chin

smaller front teeth

A comparison between a Neanderthal skull found at La Chapelle-aux-Saints (*top*), and that of an anatomically modern human from Cro-Magnon (*above*), both sites in France, with key characteristic features labelled.

One other site of note from this early Neanderthal period is Pontnewydd Cave in Wales. A total of nineteen teeth from children and adults have been recovered from this site and dated to about 230,000 years ago. They show a feature called taurodontism, where the roots of the molars do not branch in two as distinctly as in typical modern molars, but instead have a large single 'pulp' space for more of their volume under the gum line. Taurodontism is typical for Neanderthal molars, but in modern humans is a rare condition considered to be an abnormality.

The Pontnewydd Neanderthals are notable for being not only among the first Neanderthals in Europe, but among the last in Britain for more than 100,000 years. According to researchers from the Ancient Human Occupation of Britain project, a major collaborative research team spanning a number of institutions, there seems to be a hiatus in the human occupation of Britain from about 200,000 years ago to 60,000 years ago. They associate this with the formation of the English Channel as a barrier, leading to the gradual extinction of the local early Neanderthal population in Britain. When the land area between Britain and France re-emerged during a subsequent glaciation, Neanderthals finally returned.

Channel Island barbecue

In 2008 an amateur archaeologist from the Netherlands called Jan Meulmeester found twenty-eight handaxes, along with mammoth bones, in a sand pile left by a company dredging the English Channel. It is hard to imagine a more potent reminder that this seaway was once a land bridge. Yet one does exist.

Jersey, some 22 kilometres (14 miles) off the coast of Normandy, is the largest of the Channel Islands. Known mainly as an offshore tax haven, it contains one of the most impressive Neanderthal sites in Europe. Neanderthals returned here time and again for most of the period covered in this and the next chapter, when Jersey remained connected to France even as Great Britain was becoming cut off from the continent. What drew them to this place, which, fortuitously for us, is still above sea level, is that the conditions were perfect for hunting large game.

In 1910 the Oxford anthropologist Robert Marett started excavating at a dramatic seaside cliff on Jersey known as La Cotte de St Brelade (see plate 7). He found what turned out to be Neanderthal teeth and extensive evidence of habitation. His team was the first of three to work at the site. The second team, led by the Cambridge archaeologist Charles McBurney, worked at La Cotte for most of the 1960s and 1970s, uncovering an extensive pile of mammoth and rhinoceros bones, which dated to before the Last Interglacial, 130,000 years ago.

As McBurney's work was coming to an end, his field assistant, Katharine Scott, was wrapping the animal bones in fibreglass bandages to preserve them for the trip back to Cambridge, when she looked up, the freezing rain streaming down her face, and it occurred to her that the woolly mammoth and rhinos must have fallen from the cliff top directly above her. What clinched the argument for her was the quantity of skulls on the pile. Mammoth skulls are heavy relative to their nutritional value. If the animals were killed and butchered somewhere else, why did the Neanderthals bring so many useless body parts to this spot?

Combined with the observation that the bones were concentrated closely together, Scott put forward the theory that the mammoth remains represented two separate mass-kill events. She argued that the animals were driven over the cliff to their deaths – a hunting method well documented in other parts of the world in more recent times – providing an instant, if excessively large, feast for the Neanderthal hunters.

The effort involved in driving a herd of mammoths over a cliff is truly remarkable. There is extensive evidence of Native Americans creating 'drive lanes' to funnel bison over cliffs. They achieved this with controlled grass fires and by different members of the hunting party positioning themselves at key places along the route and then bursting forth at the right moment to keep the herd moving. For Neanderthals to achieve a cliff fall like this, they would have had to choreograph and execute a complex series of moves, testifying to their ability to plan several steps ahead and communicate that plan. While La Cotte would be the earliest and biggest cliff fall site, there are other possible examples throughout Europe attributed to the Neanderthals, which involve horses, bison, reindeer and wild cattle.

After more than a century of archaeological work at La Cotte, the site has served as an inspiration for generations of Neanderthal researchers. Archaeologist Clive Gamble looked back to a visit to McBurney's office in Cambridge as a key reason he was drawn into the field: 'In his room were some of the rhino skulls, and he would explain in graphic terms how they were driven over the cliff edge and then dragged beneath the overhang. It was also the richness

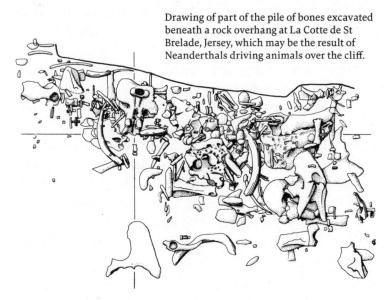

Drawing of part of the pile of bones excavated beneath a rock overhang at La Cotte de St Brelade, Jersey, which may be the result of Neanderthals driving animals over the cliff.

of the site in terms of artifacts – something quite remarkable for the British Middle Palaeolithic, a real super-site.'

In 2010 a new team resumed excavations at La Cotte with the hope of reinterpreting the site. Co-directed by Martin Bates, Chantal Conneller, Matthew Pope, Beccy Scott and Andy Shaw, this project is a partnership that spans five institutions. This latest team has been prevented on health and safety grounds from excavating under the cliff, where the bone pile is located. Their analysis of the topography above the rockshelter led them to re-evaluate the cliff fall theory because they felt a cliff fall would be impractical here. In addition, they believe that the bone pile could be interpreted as a butchery and habitation site. They see the site not as a cliff fall but as an attractive rockshelter that had commanding views over the now-submerged coastal plain.

Many of the bones at La Cotte show signs of burning. The Neanderthals' mastery of fire, which would have been necessary to survive in Europe, especially in glacial periods, was clearly established. These bones may have been burnt as part of the cooking process or as a substitute for wood. This extensive evidence of burning invokes for us the image of a mammoth barbecue.

In addition to the complex behaviour involved in hunting and cooking these large animals, one more sign of modernity comes from handedness studies. Recall that the population at Sima de los Huesos (see Chapter 3) was predominantly right-handed. An examination of how the tools were made at La Cotte has led to an estimate that 80 per cent of the toolmaking Neanderthals of Jersey were right-handed, which is close to the proportion of modern humans today. For reasons possibly related to brain hemisphere functions and dating back to the earliest members of the *Homo* line, our species has evolved to be increasingly right-handed, and at La Cotte we see that Neanderthals had paralleled our trajectory.

Pope has described La Cotte as the greatest long-term occupation site in Europe, by which he means that Neanderthals returned here consistently over many millennia, not that they lived here permanently. In addition, this is possibly the site with the longest evidence of barbecuing – spanning more than 200,000 years – in the human past. La Cotte is also remarkable for the fact that the Neanderthals imported good-quality flint and made what are called 'Levallois' flakes – which required as much forward planning as the hunt itself – resharpening them to the point of exhaustion. We now turn to these stone tools to find out what they tell us about Neanderthal cognitive abilities.

The first French Revolution

For the first 1.5 million years after the invention of the handaxe, there was little improvement in stone technology. This represents more than 80 per cent of the time since *Homo erectus* crafted the first handaxe near Lake Turkana, Kenya, 1.76 million years ago. It is impossible to conceive how long this is – how many humans decided to make a handaxe, to use it, to throw it away, and never to imagine that there might be a better way to make cutting tools.

Looking back, it is easy for us to see inefficiencies in the ways humans used handaxes. They are large and unwieldy, take a great deal of skill and time to make and are not the kind of thing you want to carry over long distances. People would spend the time to make handaxes and then kill and butcher game all in the same area, leaving everything behind that they did not eat (or wear). Handaxes, despite their size, were the equivalent of disposable razors in the early prehistoric world.

The first hints of innovation started to appear more than 400,000 years ago in northern France. The idea came in fits and starts, and there is evidence of experimentation with new techniques during a time when Europe entered the Great Interglacial, an extended period of favourable climate. From around 427,000 years ago down to 301,000 years ago Europe was mainly warm, with one short (in geological terms) 30,000-year glaciation. This is the period we reviewed in Chapter 3, when many European fossils showed signs of 'neanderthalization'.

Europe then became cold again, and when this next glacial cycle ended around 250,000 years ago, new ideas for better ways of making cutting tools had matured and taken root across the continent. Using these novel toolmaking techniques, humans were able to travel greater distances. By carrying not just their tools, but material they prepared especially for making new tools, they could go further away from flint sources, they could take longer, riskier journeys and they could meet more humans on the way.

Perhaps the most surprising thing about this innovation is how it logically follows from handaxe manufacture; it seems remarkable that no one had thought of it for 1.5 million years. And the second most surprising thing, at least for those who previously took a dim view of Neanderthal abilities, is that the Neanderthals in Europe came up with the idea all by themselves. We know this because the technology did not appear suddenly but can be traced back into the past of the same sites where it flourished 250,000 years ago. In South Africa the ancestors of *Homo sapiens* were following a similar trajectory of tool development, and in Georgia (with no associated hominin bones) there has been a more recent discovery of the local development of these tools, but the European dates are as old as the African and Asian ones.

So significant is this change in tool technology that archaeologists use it to mark the transition from the Lower Palaeolithic (the lower 'old stone age', which is the earliest period that archaeologists name) to the Middle Palaeolithic. This is the first technological transition in a series that later takes us to the Upper Palaeolithic, Neolithic ('new stone age'), Bronze Age, Iron Age and onward to historic times. In other words, for archaeologists, the emergence of Middle Palaeolithic tools was the first major step towards modern technology since the emergence of humans and of complex tool-making in Africa.

The village of Levallois, in a north-western suburb of Paris, France, is named after Nicolas-Eugène Levallois, an obscure 19th-century property developer. Now called Levallois-Perret, this has long been an industrial area and features a shuttered Citroën factory. Georges Pierre Seurat painted *A Sunday Afternoon on the Island of La Grande Jatte* there in 1884. Like Joachim Neumann, after whom the Neanderthal species is named (as we saw in Chapter 1), an accident of history linked Levallois, not to French car-making or Impressionist paintings, but to the Neanderthals' tools.

Excavations in Levallois-Perret by the River Seine in the late 19th century turned up fossils of extinct animals, such as northern species of rhinoceros and elephant, associated with a particular small, pointy type of stone tool. The great French archaeologist Gabriel de Mortillet argued in *Le Préhistorique: origine et antiquité de l'homme* (1883) that these were fundamentally different from the handaxes associated with Saint-Acheul, and he therefore distinguished between Acheulian and Levallois tool types.

The Levallois technique of toolmaking has since come to denote the first of the so-called 'prepared core' technologies that marked the Lower–Middle Palaeolithic transition in Europe and the emergence of the Neanderthals. How does one make a Levallois tool, and why does it represent such an important cognitive leap forward? And how did the Levallois technique evolve from handaxe-making?

There are two broad ways in which a knapper can organize the knapping process in order to make a cutting tool. The first is to create an end product, what would be traditionally called the tool, by knapping down a lump of rock, removing layer after layer of flakes. The second way is to knap flakes off the core and then work some of these flakes into the end products. In the first method the core is the intended tool, while in the second the core is simply a by-product of the knapping process, a lump of raw material from which useful flakes can be 'mined'. 'Levallois' refers to a special kind of flake tool whose form is predetermined by several preceding steps.

Flakes come in all shapes, sizes and forms and have varying degrees of complexity. Many of the stone tools we have already discussed were flake-based. For example, in Chapter 2 we saw how *Homo antecessor* used simple flakes to make tools from 1.2 million to around 600,000 years ago at such sites as Atapuerca, Pakefield and Happisburgh. And in Chapter 3 we saw evidence for flake-based Clactonian tools at such sites as Bilzingsleben and Vértesszöllös. Up to this point flake-based tools were less sophisticated than the core-based handaxe and could be made by inexperienced knappers. In Neanderthal Europe the Levallois revolution turned this equation on its head, and a more sophisticated type of flake-based tool industry came to dominate.

The handaxe was the first great core-based tool. To make one, the knapper removes several flakes from one

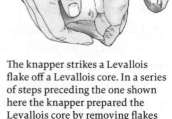

Levallois flakes from Kokkinopilos, Greece. The flake on the right has the classic, tortoise-like back of a Levallois flake (lengths approx. 53 mm and 47 mm).

The knapper strikes a Levallois flake off a Levallois core. In a series of steps preceding the one shown here the knapper prepared the Levallois core by removing flakes from the face of the core.

A Levallois point (*left*) and core (*right*) from Kokkinopilos, Greece. The core was abandoned immediately after a Levallois point (not found at the site) had been struck from it. The roughly triangular depression in the middle of the face of the core is the negative scar left from the removal of the Levallois point (lengths approx. 46 mm and 60 mm).

face, then turns the core over and removes a series of flakes from the other face, and then back over and over again. The product is symmetrical, with two flat surfaces, much like an almond, only with a sharp edge. During manufacture the knapper is not only removing flakes from one face but is also setting the stage for the removal of flakes from the opposite side.

The central innovation of the Levallois technique is that it introduced a degree of control and forward planning to making flake tools. In both the Levallois technique and handaxe manufacture, the knapper starts the same way, working down a core into a desired shape by reducing it from two sides alternately. And in both cases the knapper has to think ahead and use each flake removal to set the stage for what is to come further down the line. This is why many researchers believe that Levallois evolved organically from handaxe making. The difference in Levallois is that the two sides are not symmetrical, because only one side is the intended surface from which a predetermined flake of a desired shape and size is removed.

It is easy to get lost in the technical definition of a Levallois flake. What is perhaps of greater importance is what is left behind. In handaxe manufacture, the core at the end of the knapping process is the intended tool. In the Levallois technique, the core is essentially a tool whose purpose is the manufacture of other tools – Levallois flakes. The ability to make tools was once considered definitive of what separates humans from other animals. Now that we know that other primates and even birds use tools, this notion has changed, and it is now commonly thought that humans are defined by the ability to make tools that are designed to produce other tools. This was the function of a Levallois core, and by this definition it makes Neanderthals as essentially human as *Homo sapiens*.

The site of Cagny-la-Garenne is located in the Somme river valley, the same valley in northern France that includes some of the early handaxe sites such as Saint-Acheul and Abbeville, which contributed to the establishment of human antiquity. The area boasts a large amount of high-quality flint sources, and there was an extensive handaxe-making tradition here. It is exactly the kind of place where you might expect someone in prehistory to have started making a handaxe and then experiment by shifting to the manufacture of flake tools. In many cases such experimentation seems to be linked to a mistake where a handaxe was broken in the course of knapping. It is at Cagny-la-Garenne where the oldest Levallois flakes in Europe have been found, dating to over 400,000 years ago. Just as incipient Neanderthal traits are evident in the fossils at Sima de los Huesos at a similarly early date, we can see the beginnings of Levallois technology from that time, but only in a restricted area. It is only at around 250,000 years ago that all the trends – physical form, brain size, forward planning and the Levallois technique – come together and become widespread.

Why did this new knapping method come to dominate Neanderthal Europe, replacing the handaxe in importance? What was its competitive advantage? For one, the cutting tools are smaller and can be carried further. Also, Levallois points can be hafted onto a spear, and there is evidence that Neanderthals used red ochre for this purpose. From later periods, there is evidence that they used birch tar as an adhesive, and this requires even more steps, and therefore more forward planning.

In many cases it is possible to identify the sources of the stone that formed the raw materials for tools used at a site. We can therefore see how far away from the site the sources were located. We can see whether raw stone was brought back to the site and knapped there or whether it was knapped closer to the sources. We can also see whether tools made on certain materials were resharpened and reused extensively or were discarded after one or two rounds of use only.

For Neanderthal sites like La Cotte, it has been documented that raw material from distant sources is of higher quality than that from nearby sources. In addition, it is clear that the stone tools from distant sources came to the sites already shaped into flakes or retouched tools and that they were reused more intensively than those made from materials from nearby sources. The Levallois technique allowed Neanderthals to travel further and with better stone tools than their predecessors. The archaeologist Steven Kuhn has built on this idea, seeing in Levallois tools the ability to pre-plan resource needs and match them to particular sites, indicating that the Neanderthals were efficient hunters and food gatherers.

With Levallois tools the Neanderthals would have improved their chances of securing food, especially from migrating animals, and it would have allowed them to meet more groups, increasing their opportunities for finding sexual partners. Increased communication would also have facilitated the exchange of ideas and the consolidation of their technological advances. In sum, Levallois tools required a new level of forward planning to produce, and these tools had tremendous benefits over the handaxes and simple flakes that preceded them.

Cagny-la-Garenne and other sites like it have demonstrated that Levallois is the earliest form of prepared core toolmaking. But there were others as well. One important one is blade production. It would be natural to assume that all stone tools with a cutting surface are blades, but this is in fact a technical term that refers to elongated flakes. Blades by definition have to be at least twice as long as they are wide and must have one or more ridges running parallel to their long edges. Other than the occasional lucky accident, blades can only be made with the prepared core technique.

For many years blade production was considered to be a distinguishing characteristic of modern humans. But the current evidence is that the Neanderthals actually produced blades at a very early stage at sites on either side of the English Channel, just as the prepared core technique started to gain traction. The difference is that *Homo sapiens* developed the technology further, using more efficient production methods, and turned blades into the dominant lithic artifact of their times. Neanderthals, on the other hand, are known mainly for wider stone tools in a shape similar to the early Levallois flakes.

While Germany has laid claim to type-sites for the species *Homo heidelbergensis* and *Homo neanderthalensis*, France's contribution to archaeological nomenclature has been in stone tools: Acheulian, Levallois and Mousterian. This is in part because France was at the heart of the Neanderthal occupation of Europe, and they left many stone tools behind for French archaeologists to classify. Since the late 19th century and continuing today, the French are to lithic studies something akin to what they are to haute couture or haute cuisine.

Neanderthal tools, whether made through Levallois or non-Levallois techniques, are known as Mousterian, after the site of Le Moustier in France. Le Moustier is a rock shelter in the Dordogne where a well-preserved Neanderthal skeleton was found in the early 20th century in association with tools. Thanks to this discovery, the Mousterian has long been almost synonymous with the Neanderthals, though it was later discovered that archaic *Homo sapiens* outside Europe also made Mousterian tools that are indistinguishable. In Chapter 5 we will see how Neanderthals and early modern humans produced similar stone tools when the two populations arrived in Asia.

In a curious postscript to the Le Moustier story, told by Erik Trinkaus and Pat Shipman in *The Neandertals* (1992), the original Le Moustier skeleton was well travelled after its excavation, having been sold to the Museum für Volkerkunde in Berlin and then taken to St Petersburg as war booty. A second Neanderthal skeleton from the site, this one an infant, was lost for many decades only to be rediscovered in the early 21st century in the collections of the Musée National de Préhistoire in Les Eyzies.

Mousterian flakes and points (which have a sharp tip) look surprisingly uniform across the vast geographical and chronological range covered by Neanderthals. Apart from some idiosyncratic, localized forms, and differences in raw materials, it would be hard to distinguish a Mousterian stone tool from, say, a 200,000-year-old site in France from one in Crimea from 50,000 years ago. Thanks in part to this uniformity, the Neanderthals have been charged with limited capacity for innovation.

Under this superficial homogeneity, however, there is immense variability within the Mousterian, deriving from different ways of cutting up a core into flakes and blades. In other words, there were many ways that the Neanderthals could arrive at the same end product, and we can read patterns into which techniques were favoured at different times and places. The stone tool assemblages from some Mousterian sites contain one or several different variants of the Levallois and even other prepared core technologies. In other

Mousterian sites there is little or no use of the Levallois. These sites are no less 'Mousterian' than those with high Levallois frequencies.

Bordes versus Binford

While much of the evidence for the cognitive advances of early Neanderthals comes from stone tools, it is ironic that many of the researchers who first analysed these tools did not give Neanderthals much credit for their capabilities. Yet these researchers coined many of the terms and pioneered the analytical techniques that still dominate Mousterian lithic studies.

The French archaeologist François Bordes, one of the patriarchs of modern Palaeolithic archaeology, first brought the variability of Neanderthal tools into focus in the 1950s and 1960s. Bordes deliberately avoided saying much about Neanderthal cognitive abilities or social life. The epilogue of his book, *A Tale of Two Caves* (1972), an English-language semi-popular account of his decade-long work at two Mousterian caves, Pech de l'Azé and Combe-Grenal in the Dordogne, starts with these sentences: 'Perhaps this tale of two caves has proved a disappointment to you. Perhaps it contained too much cold science and not enough about the human life of those remote times. But, unless I wish to write a science fiction story, it is not possible to go further.' Using the pseudonym Francis Carsac, Bordes was, in fact, also a prolific science fiction writer. In his 1955 novel *Les Robinsons du Cosmos*, for example, an entire French village is teleported to a faraway planet. Despite his reluctance to address directly the question of Neanderthal cognition, he clearly had low opinions of it.

While his predecessors relied on the mere presence or absence of some characteristic tool types to draw conclusions, Bordes examined the entire lithic collection recovered from each site. He achieved this by dividing and subdividing Mousterian tools to the point of exhaustion. He defined fifty-nine different types of modified,

STONE TOOLS: THE BASICS

There is an irony in comparing the small amount of time our predecessors spent in making stone tools with the endless hours Palaeolithic archaeologists (one of the authors of this book included) invest in measuring, drawing, recording and analysing them. The ancients at least got food and prestige for their efforts. What have we gained from ours? For those new to lithic analysis, we offer a brief introduction to the subdiscipline of archaeology that gives perhaps the best insight into the ways that Neanderthals thought.

Stone tools are the most common type of artifact from Palaeolithic sites. Organic remains, such as animal bones, tools made of bone or antler, and seeds or other plant remains, are less likely to survive. Even rarer are human fossils, and tools or containers made of wood or reeds. Most archaeological sites have not yielded any fossils at all, whether human or animal.

Knapping stone tools may look to the uninitiated like little more than banging rocks together. In fact it is a controlled and predictable process, governed by the laws of fracture mechanics. If you have ever been heartbroken yet intrigued by the way chips of glass break off an otherwise lovely crystal bowl or coffee table, you have the basic idea. One applies force to a lump of rock usually by hitting it with a stone cobble called a hammerstone. A crack will develop and spread from the

A Mousterian side scraper from Morfi, Greece (length approx. 93 mm).

point where the hammerstone strikes the rock. Given enough force, the rock fractures into two or more pieces.

Only certain kinds of stone can be used for knapping stone tools. It needs to fracture easily and predictably. The smoother the better, which is why the most elaborately knapped material that archaeologists have found is the volcanic glass called obsidian. The slightly less fine-grained and less homogeneous flint and quartz are more widely available in nature and were the main materials used in prehistoric (and historic) times for knapped stone tools. More coarse-grained materials have also been used. We tend to associate these with the simpler stone tools made by the earliest of our ancestors, but they were used throughout prehistory. In areas where good raw materials were not readily available our ancestors have even knapped seashells, limestone or wood into tools.

If the knapper knows how to control the angle and force of the blow and the exact point where the hammerstone hits the rock, he/she can control the shape and size of the pieces that come off the lump of rock (the flakes) and the shape of what remains of the lump of rock (the core). By controlling the shape of both the flakes and the core, the knapper can string together a sequence of strikes to the core to remove flakes, which results in a complex end product.

Humans first used hard hammers (hammerstones) for knapping. They started to use soft hammers (antler, wood or bone) perhaps 700,000 years ago for finer control. Pressure flaking, where soft hammers are pushed (not struck) to cause a fracture, came along in the Upper Palaeolithic, after the time of the Neanderthals.

The flakes or blades that come off the core are sometimes reworked (or, in lithic terminology, 'retouched'). In this reworking, tiny flakes are knapped off a flake's edges in order to give it a desired shape (such as straight or curved) or a particular morphology (such as serrated or pointed). Retouching can also be used to resharpen tools that have become dull from use. In the course of their useful lives, lithic artifacts can be potentially reworked into different forms with different functions. Unlike in most other technologies, a lithic artifact is essentially never finished.

Traditionally a 'tool' was considered to be either the desired end product of knapping or a retouched flake. In the last few decades, new techniques to examine microscopic wear marks and residues left on the surface of used stone artifacts have demonstrated that flakes that were not retouched at all were also used regularly. In fact, when a flake is struck off the core its edges are fresh and very sharp (often sharper and thinner than a modern surgical scalpel).

How can archaeologists recognize lithic artifacts? In knapping, when the hammerstone hits the rock, the force spreads through the material in concentric shock waves. These shock waves are well pronounced and close to one another near the point of impact and become less marked the further away they travel from the point of impact. Think of the ripples created when a pebble hits the surface of a pond. If the knapped lump of rock splits apart in two pieces, each of these two pieces will have a new, slightly curved surface with concentric ripple marks reminiscent of the ripples of a shell. (Hence the formal name for this type of fracture, 'conchoidal', derived from the Greek word for shell.) These two rippled surfaces are each other's perfect negative. At the point where the hammerstone hits the rock, the flake has a half-conical bulge, which is called the bulb of percussion, while the same spot on the core retains the negative of the bulb.

This 'conchoidal' type of fracture, with the bulb and ripple marks that it leaves on the knapped pieces of stone, is the basis of the archaeological analysis of stone tools. At the most elementary level, it provides a way to distinguish stone tools (rocks intentionally knapped by humans) from similar-looking rocks fractured by natural processes such as rockslides. This is particularly useful in early sites, where the tools found are often just simple flakes and may not be accompanied by any hominin fossils or other corroborating finds. Indeed, there are often controversies over the human role in shaping the stones at such sites.

Flint-knapping is essentially a process of peeling material off a lump of rock, layer by layer, much like peeling an onion. Based on the scars and the sets of ripple marks left on a core, it is possible to reconstruct how each successive 'layer' of material was 'peeled' off the core.

The back of each flake also bears the partial scars left on the core by earlier flake removals. To the lithics specialist, each flake and each core tells a little segmented story of how it was made. Putting together all the little stories from the tens or even thousands of stone tools recovered from a site, one can reconstruct the bigger story of how the stone tools of that site were made.

As the flint-knapper works down a core, removing flakes layer by layer, each piece that falls off the core remains on the ground, around where the knapper sits. In some cases it is even possible to reconstruct where a knapper was positioned at a site. Often the two rippled surfaces can be fitted back together (or 'refitted' in lithic analysis lingo). Refit all or most of the flakes removed from the same core and you end up with the original lump of rock in a three-dimensional jigsaw puzzle. This has been done successfully at many sites, notably Boxgrove and Maastricht-Belvédère.

It is no surprise that of the two authors of this book, the lithics specialist is also a jigsaw puzzle aficionado. But this is not just a complicated mental game, befitting the modern human brain. Through refitting and experimental replication of flint-knapping we can come as close as we can to seeing our predecessors' thought processes in action.

The choices of the toolmaker do not start when the knapper sits down to start knapping and do not end when the end product has been successfully made. The knapper chooses a lump of raw material appropriate for what he/she wants to make on each particular occasion. Going even further back,

A Levallois blade from Morfi, Greece. The left edge retains a thin sliver of the rough external surface of the lump of stone from which the tool was knapped. The knapper did that intentionally, as the rough surface provided a ready-made blunt edge to hold the tool. In Bordes' typology, this is a naturally backed knife. The right edge was the sharp cutting edge (length approx. 78 mm).

A refitted core from Maastricht-Belvédère, Netherlands. The flakes removed from a single Levallois core have been painstakingly fitted back together.

the knapper chooses what raw materials to acquire, what sources to acquire them from, how much time and energy to invest in acquiring good raw materials, whether to do the knapping at the raw material source or at a 'home base' site or a hunting site. Then, before each strike, the knapper is presented with a situation to assess and with a number of alternative ways to proceed.

In some Palaeolithic sites we can reconstruct knapping 'journeys' in enough detail that we can see at each step what situation the knapper was faced with, what were the alternative courses of action available and what was the action chosen by the knapper.

All these choices are linked to when, where and for what activities the stone tools will be used, resharpened, reused and ultimately discarded. The different stages in the 'life' of a stone tool (raw material acquisition, knapping, use and discard) are interlinked, and at each of these stages our predecessors made choices that required planning, envisioning far ahead what the repercussions of these choices might be.

retouched tools, plus four types of unmodified flakes and points produced by the prepared core technique. He then used simple statistics to understand which of these tool types was most prominent in each assemblage.

Of Bordes's sixty-three tool types, twenty-one are different varieties of 'side scrapers', which are flakes or blades on which one or two long edges have continuous retouch along the whole length of the edge. Despite the name, side scrapers would also have been useful for cutting. Bordes's near-obsessive categorizing highlights the perils of trying to name an artifact type after its supposed purpose.

When Bordes looked at the relative abundance of side scrapers and the other main tool types in Mousterian assemblages from sites in northern and south-western France, he realized that they fell into distinct groups: the Charentian Mousterian (dominated by side scrapers), which he subdivided into the Ferrassie group (in which the Levallois was often used) and the Quina group (in which Levallois products were rare); the Denticulate Mousterian (dominated by tools with serrated edges which he called denticulates and notches); the Typical Mousterian (in which both side scrapers and denticulates and notches are common, in roughly equal measures); and the Mousterian of Acheulian Tradition (in which there are handaxes and backed knives, which are blunted to make them easier to hold, in addition to abundant side scrapers and denticulates). Subsequent research since the 1960s elsewhere in Europe and the Near East has supported Bordes's findings that these are indeed distinct groupings within the Mousterian.

In *A Tale of Two Caves* Bordes interpreted these different variants of Mousterian lithic industries as 'different ways of performing the same activities with different tool kits'. For him the roots of this variability were cultural. People were taught specific ways of doing things and continued to use them and pass them down the generations: 'Our point of view is that during Mousterian times different cultures, with different traditions of toolmaking and tool-using, coexisted on the same territory but influenced each other very little.'

Bordes actually used the names of his groups of lithic industries to refer to the makers of these industries as distinct, almost ethnic groups (such as the 'Typical Mousterians', the 'Denticulate Mousterians'). Stone tools become people: 'Intermarriages are difficult to assert or refute, but in primitive societies, conservatism is usually very strong. If one supposes that a Mousterian of Acheulian Tradition man married a Quina woman, she might have gone on using the thick scrapers to which she was accustomed, but we doubt that her daughter would have done the same.' There is little if any room here for change, innovation and spread of new ideas.

The strongest rebuttal to Bordes's interpretation came from across the Atlantic. Lewis Binford, a larger-than-life, almost messianic figure credited with launching the 'New Archaeology' of the 1960s, argued in a paper with Sally Binford that the divisions Bordes identified were functional rather than cultural.While Lewis Binford remains more famous than Sally, it was Sally who did most of the writing in their seminal paper. Sally – who met Francois Bordes and his wife Denise de Sonneville Bordes in 1960 – was also the one who was familiar with the French material. She published a number of important papers with Lewis in the late 1960s but divorced him and left archaeology in 1969 to become an activist for social causes.

In their paper, the Binfords claimed that the different variants of Mousterian industries were made not by different groups of Neanderthals, but by the same groups while working on different activities. Certain tools were used together as distinct tool kits, each for specific functions and activities only. The activities for each site depended on the distribution of natural resources across the landscape. The Binfords did not elaborate much on the composition and uses of these tool kits. The two tool kits that they referred to in detail were scrapers and points on the one hand, and the denticulates and notches on the other. They suggested that scrapers and points were used for butchering and processing animals, while denticulates and notches were used for processing plants, wood and possibly cracking open skulls and long bones to access the brains and bone marrow.

The Binfords went further and suggested that the primary processing of animals was done mainly by the males of the group, while the processing of plants or bones was done by the females. In other words, Bordes's 'Quina woman' was replaced by the Binfords' 'denticulate woman' who made only denticulates and notches (not the more elaborate scrapers and points) and only did the humble work of processing plants or cracking open bones.

The Binfords' argument was part of a distinct view of Neanderthal society, a view that for some time was very influential, at least within the Anglo-Saxon world of Neanderthal archaeology: Neanderthal men and women led fairly separate lives, with separate subsistence and other activities and little inter-group cooperation. Neanderthal life was expedient and day-to-day with little forward planning, mostly responding to immediate needs and availability of resources.

We now know that the Binfords were wrong on most points. They considered that Mousterian tools were made, used and discarded on the spot, but in fact some were transported far from where they were made. They thought the Neanderthals mostly scavenged dead animals or killed the weaker young, old or diseased animals, but in fact they did hunt animals in their prime. And the Binfords thought that the lithic points they made could not function as projectile tools, but studies have shown that they could.

Despite these faults, the Binfords' approach had an enormous impact. According to Clive Gamble, the Lewis and Sally Binford paper on the Mousterian 'was a game changer in more ways than one since it tried to do something with the stone tools other than describe and typologize them. Looking back it's difficult to remember how contentious it was. However, so much flows from their paper in terms of a landscape archaeology for the Neanderthals that its influence on me was immediate and long-lasting.'

François Bordes and Lewis Binford, along with their wives who were their research partners, shared a goal of making Neanderthal archaeology more scientific, and they relied heavily on statistics. They remain influential for Bordes's notion of analysing an entire lithic assemblage and for the Binfords' emphasis on function. More recently, archaeologists such as Harold Dibble have built on their work by demonstrating that many of these different tool types can be products of a longer chain (for example, through frequent use and resharpening, single side scrapers can evolve into double side scrapers and then convergent side scrapers). Where Bordes and the Binfords fell short was that their theories did not credit the Neanderthals with the ability to innovate. The Neanderthals were far more advanced than they imagined.

Neanderthal society

Using the evidence of stone tools and bones, which is almost all we have from this distant epoch, what can we reasonably say about Neanderthal social life? By applying the concept of forward planning to social units, we can posit that Neanderthals were better than their ancestors at anticipating what their friends and family might do in different situations. In other words, they were becoming increasingly sophisticated, enabling them to create stronger bonds and larger social units, ultimately improving their survival abilities.

This line of reasoning follows the social brain hypothesis, which we mentioned earlier in the chapter. The key insight of the hypothesis is that our species is a highly social one and that group size is a major part of our success. The intoxicating feeling of being part of a large group is something all of us experience, whether it is singing or cheering in unison at a sporting event or praying together in a religious setting. In daily life, humans not only crave this feeling of group belonging, but we depend on the cohesion of groups for our survival and advancement. A 'group' is not necessarily a collection of people that live together, but is perhaps better described as a network of relationships. A group is made up of people we recognize by sight or voice as being reliable friends or allies.

According to the social brain hypothesis, the increasing complexity of social skills needed to navigate human groups was probably the main evolutionary driver behind the increase in brain size that we see from the earliest *Homo*. Essentially, it was a positive feedback loop. As social groups became larger, more cognitive processing power was needed for an individual to succeed, which in turn led to even higher complexity. The challenge of complex social relations was a factor for both Neanderthals and archaic *Homo sapiens* – which may explain why both populations continued to evolve larger brains, and with them larger group sizes, in parallel.

The author of the social brain hypothesis, Robin Dunbar, has related social group size to the size of the neocortex, which is the technical term for the outer layers of grey matter in the brain. The neocortex is the centre of higher order brain functions such as visual and auditory perception, social and emotional processing, language, memory and learning. Here is where size is not everything in brains, for modern humans have a skull that is dome-shaped, allowing for a larger neocortex than the flatter Neanderthal skull, despite the fact that Neanderthal brains were as large as or larger in total volume than our modern brains.

According to Dunbar, chimpanzees, probably like our common ancestor with them, live in groups of fifty to fifty-five individuals. Modern humans, he argues, have brains that can maintain social relationships with 150 others, the so-called 'Dunbar number' which we can see across many aspects of human activity such as effective unit sizes in the military, sizes of companies or corporate departments, and in social networking.

Chimpanzees maintain group cohesion with grooming. But grooming is time consuming, giving a natural limit to the number of individuals with whom different group members can interact. We maintain cohesion through small talk, which is much quicker, allowing for bigger effective groups. Somewhere between our common ancestor with chimpanzees (around 6 to 8 million years ago) and the present, our ancestors developed language, enabling a steep increase in group size. Dunbar places this moment at about 500,000 years ago, which gives credit to *Homo heidelbergensis* and its contemporaries for the development of complex speech. Genetic studies, particularly of the FOXP2 gene, which we discuss in Chapter 6, place this development at least that far back, and perhaps even earlier. The idea that speech developed so early is also supported by a modern-looking hyoid bone, essential for speaking, found in a Neanderthal at Kebara Cave in Israel.

Chimpanzees use grooming to show friendship and promote social cohesion. The social brain hypothesis states that at some point in our evolutionary past, the introduction of speech (small talk) achieved these functions more efficiently than grooming and allowed for larger social groups.

Using Dunbar's theory, we can build up a picture of Neanderthal social life, which was probably quite similar to what our own ancestors were doing in Africa during this period. Neanderthals probably had social networks slightly smaller than ours, perhaps of 120 individuals, most likely structured along the lines of modern hunter-gatherers, who are based in clusters of about thirty people (four to six families) and have good relations with neighbouring clusters. We should note that some researchers think that Neanderthals lived in even smaller clusters, judging by the small sizes of some of the caves and rock shelters they inhabited. One of Dunbar's students, Eiluned Pearce, argues that Neanderthals had larger eye sockets than modern humans, indicating that Neanderthal brains gave more space to visual processing at the expense of social cognition. Based on patterns of sexual dimorphism, Dunbar thinks that Neanderthals practised polygyny, with two women for each man. Curiously, these theories rarely seem to make it into museum displays and illustrations, which often emphasize anachronistic monogamous nuclear families.

Having become adept at forward planning, using the Levallois technique to make stone tools and organizing themselves enough to hunt large animals in their prime, the Neanderthals were well positioned to take further leaps forward. Yet most of the period from 250,000 to 130,000 years ago had a climate of sustained cold conditions. Survival became the priority for them in the northern latitudes of Europe. When this difficult era ended, the Earth entered the Last Interglacial or Eemian, a 10,000-year-long warm period that was as mild as the planet would be until the current Holocene. It was at this time that the Neanderthals first ranged outside Europe. When they ventured forth and reached Asia, they were not alone.

An End to Isolation

130,000 to 60,000 years ago

On a clear day it is possible to see the 14 kilometres (9 miles) across the Strait of Gibraltar, which separates Spain from Morocco, Europe from Africa and – more poignantly for our purposes – Neanderthal territory from modern human territory. Around 600,000 years ago, *Homo heidelbergensis* had reached Europe, probably from Africa. For hundreds of thousands of years following this migration, the European and African lines of *Homo*, separated by the Mediterranean, evolved on separate paths. Both were developing in parallel, with their respective brain sizes growing ever larger and their behaviours becoming more complex. Yet their appearances were unmistakably and increasingly distinct.

By 130,000 years ago, at the start of the long and mild Eemian interglacial, Neanderthals occupied the land north of the Strait of Gibraltar at sites such as Gorham's Cave, Gibraltar, while *Homo sapiens* lay to the south, at sites such as Cueva Benzu, Morocco. Here were two human species on either side of a narrow channel, almost able to see each other but probably unaware of each other's existence. Both species would soon expand into Asia, ending hundreds of millennia of geographic isolation.

We know how this story ends, for it is we who are telling it. While there are many questions that are as yet unanswered about how we got from the beginning of the Eemian to where we are now, with more than seven billion living *Homo sapiens* and no other

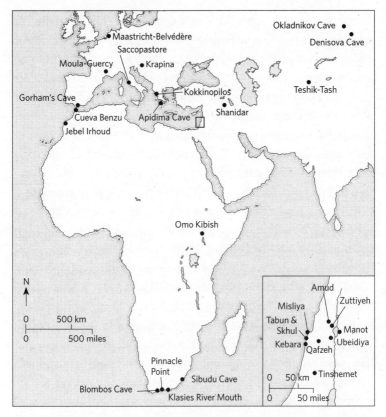

Map showing Neanderthal and early modern human sites discussed in this chapter, with an inset showing sites in Israel.

human species on the planet, the sequence of events is coming into focus. Recent discoveries and new scientific techniques have brought some real surprises. Since the 1980s our understanding of these first stages of global expansion by Neanderthals and modern humans has been turned completely upside down. In this chapter we look at the initial forays into Asia and reveal which species achieved a more stable presence there, pushing ever deeper into the continent. Hint: it's not who you think.

In the previous chapter we looked at how Neanderthals and archaic *Homo sapiens* increased their forward-planning abilities, which we can see in the way they made stone tools and hunted. During the period we cover in the present chapter, another modern behaviour emerged in both species: burial of the dead. Back in the days of *heidelbergensis*, as we saw in Chapter 3, there was a collection of human remains that reached the bottom of the Sima de los Huesos at Atapuerca. These individuals arrived in what may be the world's first mass grave either through intentional deposition (i.e. disposal down a long shaft) or through abandonment of corpses in a cave, from which they were later washed down into the deep pit. However it happened, the behaviour at Atapuerca was not burial, which involves the intentional digging of a grave and then the placing of a body (and sometimes burial goods) inside it.

There is evidence of mourning and attachment to a dead relative (especially a child) from many different species, even non-primates such as hippos, who have been known to wail in the face of loss. But intentional burial, especially within sheltered living spaces like caves, implies a more sophisticated understanding of death, perhaps even a belief in the soul. Burial demonstrates respect for the individual beyond life and the notion that if the dead are treated well, laid to rest as if in a permanent sleep, then their memory will live on, enabling them to continue to provide comfort to the living. Dealing with death is a life-cycle event that binds us together and reminds us of the sanctity of life and our shared humanity.

Around 100,000 years ago this particular human practice extended to both *Homo sapiens* and *Homo neanderthalensis*. It is not possible to say whether the oldest evidence of burial belongs to modern humans or Neanderthals, thanks to uncertainties in the dating. What is agreed is that both species started the practice in caves and rock shelters at roughly the same time and that the practice persisted.

By the end of this chapter our narrative will bring us well within the last hundred millennia. Were any differences yet evident between Neanderthals and modern humans that would indicate which variety of human might prevail? The answer is yes, but the differences were subtle. Modern humans were beginning to develop an advantage, not in hunting tools or muscle power, and not in overall brain size, but in ornamentation in the form of paint and jewelry. Our ancestors were exploring forms of symbolic expression, while there is limited evidence that the Neanderthals had taken this step. Was this the key to our success?

Asian arrival

Much of what we know about Neanderthals and modern humans in this period comes from a cluster of sites in northern Israel. In this tiny slice of Asia there are good sources of fresh water, a large number of caves, dry conditions and soil conducive to bone preservation. We are doubly fortunate that both species started intentionally burying their dead in caves at this time, making it easier for us to find complete skeletons, and also that there has already been around a century of Palaeolithic research in the area.

It is unusual for such a tiny region to be the main source of evidence for developments across a whole continent. This is akin to studying the changing population of a city over time by looking at the apartments on a single floor of a building. But this particular sample area has given us a tremendous amount of information. Remains of more than forty-five individuals from 130,000 to 60,000 years ago have been excavated from at least six caves or rock shelters in northern Israel. This is a truly staggering total from such a small area.

Most of the key hominin sites in northern Israel were discovered in the 1920s and 1930s, and much of the early work was done by two British archaeologists, Dorothy Garrod and Francis Turville-Petre. Both had gone to Oxford in 1921 to study under

Robert Marett, who had earlier found Neanderthal teeth at La Cotte de St Brelade on Jersey.

Soon after his graduation, Turville-Petre joined a project through the British School of Archaeology in Jerusalem in what was then the British Mandate of Palestine. At just twenty-four years of age in 1925, Turville-Petre found 'Galilee Man', the name given to a skull he unearthed at a site called Zuttiyeh Cave. Debate continues about the species represented by 'Galilee Man'. Estimated to be more than 250,000 years old, it probably pre-dates the emergence of Neanderthals and *Homo sapiens*. It seems to have various characteristics in common with both of these species and probably descends from *Homo heidelbergensis* as a regional variant – basically a cousin of the Neanderthals and perhaps tied to the enigmatic Denisovans, a population originally identified through DNA extracted from bones in a single site in Siberia. In 2021, nearly a century after the discovery in Zuttiyeh Cave, skull fragments and a lower jaw with similar archaic features were found in Nesher Ramla, Israel. The excavators from the Hebrew University of Jerusalem date this to around 126,000 years ago, which makes it too late to be an ancestor of either Neanderthals or modern humans. Together, 'Galilee Man' and the Nesher Ramla *Homo* remind us that there is more variation than can be explained by neat categories.

The first evidence that Neanderthals and *Homo sapiens* (which are called early modern humans, because they were not yet fully modern in appearance) reached this small slice of northern Israel began to emerge a few years after Turville-Petre's first discovery. In 1927 quarrying activity on Mt Carmel near Haifa turned up more sites, and Garrod began a multi-year excavation project in 1929. Garrod was older and more experienced than Turville-Petre (her archaeology diploma from Oxford had come some years after she studied history at Cambridge). By this time she had already dug in Britain, France, Gibraltar and Iraq. Garrod excavated at Skhul

Francis Turville-Petre at the site of Zuttiyeh Cave in modern Israel, where he discovered 'Galilee Man' in 1925.

and Tabun, two caves on Mt Carmel, and started to find a trove of ancient human remains. Despite their close proximity to each other, Skhul contained what appeared to be *Homo sapiens*, while Tabun had bones with decidedly Neanderthal features. Garrod soon expanded her operations to Kebara, another cave in the area, and invited Turville-Petre to help her. Kebara has produced some very important Neanderthal remains. In 1938, just after the conclusion of her extraordinary work on Mt Carmel, Garrod was elected to the Disney Chair in Archaeology at Cambridge and became the first female professor in any subject at the ancient university, almost ten years before women attained the same status and rights as men there. Turville-Petre, meanwhile, left archaeology, travelled widely throughout Europe and rented a Greek island.

At the same time that Garrod and Turville-Petre were working on Mt Carmel, a Frenchman called René Neuville was digging at

Archaeologists revisit Kebara Cave, Israel. Dorothy Garrod began work in this cave in the early 20th century. In 1982, Ofer Bar-Yosef and his team discovered an adult Neanderthal that had been deliberately buried.

Qafzeh rock shelter near Galilee. This proved to be just as productive a site as the others. It is often uttered in the same breath as Skhul, for it was home to the other great collection of early modern human bones from the region. Amud Cave (see plate 6), which is close to Zuttiyeh, was first excavated by Hisashi Suzuki of the University of Tokyo in the late 1950s and early 1960s. Amud produced remains of some sixteen Neanderthal individuals, one of whom had the largest brain capacity (1740 ml, 60 oz) of any ancient human ever found. More recently, an excavation project at Tinshemet Cave in central Israel, conducted by Yossi Zaidner from the Hebrew University in Jerusalem since 2016, has unearthed a number of burials, but their species is not yet known.

For the first time since the two human branches separated, we have a single region that contained remains of rival populations of Neanderthals and modern humans. It is here that we must look to try to answer two of our most pressing questions: who arrived in Asia first, and which of the two species was able to maintain a presence there? Along the way, we will begin to look at the question of whether there is evidence for interbreeding, or indeed any contact at all between the populations.

How can we judge which species was the first to reach Asia, or at least northern Israel? Since the pioneering work of the 1920s and 1930s, different methods of answering the question have been overtaken by more advanced dating technology. It was only in the 1980s that archaeologists had access to dating techniques that could give a reliable answer to the question. When the answer finally came, it did not just overturn the previous answer, but it signalled the end of an entire way of viewing human evolution.

Before the 1950s there were no absolute dating techniques that covered the Middle Palaeolithic era. And, unfortunately, there was no way to compare the stratigraphy of the different sites in the region to see whether the bones in one were older than the bones at another. So archaeologists fell back on the only tool left to them, which was the notion that stone tools that look older are older (according to an established sequence from France), and that skulls that look more archaic are more archaic than skulls that share more traits with today's living humans. In other words, they simply fitted the evidence into pre-existing theories of how the evidence fitted together. They tried to come up with plausible chronologies. With the use of today's dating techniques, we now know that what seemed implausible to this first group of archaeologists turned out to be the correct interpretation.

Many of these early researchers believed that the Neanderthal site of Tabun was older than the early modern human site of Skhul, just yards away, because they believed that Neanderthals in the region were generally older than the *Homo sapiens*. Others argued that the sites dated to the same time, and the Neanderthals and modern humans were part of one mixed and highly variable population. Still others saw these sites as evidence that the Neanderthals were not an isolated European species but represented a global phase in the evolution from *Homo erectus* to *Homo sapiens*.

It is entertaining, in a way, to look back to these debates in the early 20th century with the knowledge that everybody was wrong. Intriguingly, the development of the first radiometric absolute dating technique in the 1950s did little to expose the flaws of their theories. Carbon dating compares the ratio of regular ^{12}C (carbon atoms with six protons and six neutrons) and the radioactive isotope ^{14}C (carbon atoms with six protons and eight neutrons), which decays into ^{12}C at a known rate. By seeing how much ^{14}C is left, proportionally, in organic remains, one can discern how long the carbon has been undergoing radioactive decay and therefore how long ago the organism died (and stopped absorbing new carbon).

The key problem with carbon dating is that it is less and less reliable for material that approaches 40,000 years old. Before recent technological advances with carbon accelerators and the introduction of ultrafiltration, 40,000 years ago was the ^{14}C limit, and older material gave the same signal no matter how much older it was. Coincidentally, 40,000 years ago used to be around the accepted time for the arrival of early modern humans in the Middle East. It turns out that the arrival came far earlier, but this was not possible to demonstrate with carbon dating. Instead, many researchers took a carbon-dated minimum age to be representative of a true age. This is like saying that if your mother must have been at least sixteen years old when she gave birth to you, then she was probably sixteen years old at the time.

Tabun and Skhul were put in a sequence, with Tabun at some 60,000 years old and Skhul at the carbon limit of 40,000 years. This had the benefit, at least, of fitting into various global ideas of human evolution. It could accommodate both the theory that

Neanderthals evolved into modern humans and the theory that modern humans replaced the Neanderthals.

The first hint that all this might be wrong came from Chris Stringer's PhD dissertation. Through a mathematical comparison of several different skull attributes, Stringer began to argue in the 1970s that the modern humans at Skhul were not related to the Neanderthals at Tabun and Amud, but instead were linked to two skulls that had been found in Africa in the 1960s. The second hint came from archaeologist Ofer Bar-Yosef's projects in the 1980s when archaeozoologist Eitan Tchernov studied the microfauna of several sites in northern Israel. Like the 'vole clock' used to date the first sites in Britain, microfauna (small animals, such as rodents) evolve at a fast enough rate that the species present in an archaeological layer can be indicative of age. Tchernov and Bar-Yosef started to argue that the accepted sequence of the sites was incompatible with the microfaunal evidence.

The revolutionary moment came in 1989 thanks to the introduction of new absolute dating techniques. Just a year earlier *Newsweek* broke its circulation records when it printed a cover story entitled 'The Search for Adam & Eve' about a theory that all living humans have a recent (100,000–200,000-year-old) common ancestor who lived in Africa. For most researchers, the genetic Out of Africa theory marked the end of the notion that modern humans could have evolved out of Neanderthals or that *Homo erectus* evolved into *Homo sapiens* all over Eurasia and Africa at the same time. In 1989 the scientific dating of sites in northern Israel provided corroboration of the Out of Africa theory.

What were the new dating techniques? There were three. U-series dating is like carbon dating in that it compares ratios of the unstable uranium isotopes, ^{235}U and ^{238}U, with the elements into which they decay, although U-series is used on inorganic material such as stalagmites. Thermoluminescence, or TL dating, looks at how many free electrons have accumulated over time within little traps in material such as flint or quartz. If flint was burnt in a fire or if the quartz was exposed to the sun for long enough and then buried, the electrons would have been knocked out, effectively setting the 'clock' to zero, so that the amount of electrons that have been trapped tells us the amount of time that has since passed. Burnt flint is easy to recognize, as the process leaves tell-tale pockmarks and can change the stone's colour. And finally electron spin resonance, or ESR, works on the same principle as TL and is particularly useful when applied to ancient tooth enamel.

In the 1980s these techniques were new and experimental. Archaeologists decided to use them all in order to confirm their results, and in fact the results of the U-series, TL, ESR and microfauna were all in general agreement: Qafzeh and Skhul could no longer be considered the most recent of the Israeli sites. Instead, it appeared that they were the *oldest*. ESR and TL both place the early modern

A modern human skull from Skhul, Israel, probably a little older than the Qafzeh skull. With the development of new dating techniques in the 1980s, the modern humans from Qafzeh and Skhul were redated to before the time that Neanderthals occupied nearby sites like Amud and Kebara.

A modern human skull found at Qafzeh, Israel, dated to around 100,000 years ago.

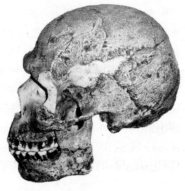

humans of Skhul between 100,000 and 135,000 years ago. Those of Qafzeh were dated by the same methods to be 90,000 to 120,000 years old. By contrast, the Neanderthals of Amud and Kebara turned out to be just about half of these ages (50,000–60,000 years old).

Researchers were surprised to discover that the first modern humans reached northern Israel, and probably the Asian continent, much earlier than the Neanderthals did. This was the one possibility that no one entertained in the days before absolute dating. It stands as an important lesson in science: in the absence of evidence, one must always be wary of drawing conclusions based on expectations, no matter how obvious they may seem. The human past seems to have a ceaseless ability to surprise us.

The reason it is important to try to use more than one absolute dating method is that individual techniques are not always reliable. Carbon dating has gone through successive stages of calibration and correction since the 1950s as researchers became more aware of its limitations and learned to compensate for contamination and for changes in the atmospheric concentration of ^{14}C. TL and ESR dating have also been improved since the 1980s. It is always instructive when different techniques (or even the same technique) produce different results on the same material, and there has been a general trend towards testing smaller sample sizes in order to eliminate contamination.

The one site we have not mentioned in this respect – Tabun – forces us to qualify the extraordinary claim about *Homo sapiens* pre-dating Neanderthals in Asia. Some radiometric dates for Tabun place the site at the same time as Qafzeh and Skhul, if not earlier. The key here is a buried skeleton labelled Tabun C1, which was dated to 100,000–120,000 years ago by ESR, while TL puts it at more than 160,000 years old. The latest U-series dates, however, suggest that it may be as recent as 20,000 years old, which, in a highly unlikely twist, would make it the last living Neanderthal identified anywhere in the world. It is possible that when the Neanderthals used Tabun as a burial site, they dug into deeper stratigraphic layers before depositing the bones, giving them the appearance of older age. Unfortunately the dating of Tabun is difficult to resolve, especially because the excavation did not benefit from modern excavation techniques, occurred before these dating techniques were developed and because the site has been completely dug out, precluding any re-examination of the layers.

We are left with some uncertainty whether the earliest humans to reach the Levant were the early modern humans or Neanderthals. In 2019, a modern human jawbone from Misliya Cave in Mt Carmel (close to Tabun and Skhul) was dated to at least 177,000 years ago. This makes it older than Qafzeh, Skhul, and Tabun, supporting the prevailing pattern of a modern human occupation giving way to a Neanderthal occupation. What it demonstrates is that humans did not evolve – as many used to believe – as a single global population undergoing a series of changes from the earliest *Homo* to the present. Instead we have a paradigm of diversity, in which there were many human species sharing the planet at any given stage until very recently.

The only other place in the world where this pattern is evident is at the site of Apidima Cave, Greece. There, researchers in 2019 used U-series dating to argue that a modern human skull is 210,000 years old, earlier than any modern human remains in Asia. The skull was found there in the 1970s, along with a Neanderthal skull. U-series dating put the Neanderthal individual at 170,000 years old. Both skulls seem to have been pushed to a higher position in the cave by rising waters and then were encased in concrete sediments. These dates have not yet been confirmed by other dating techniques and must be seen as provisional for now.

Aside from the question of which species arrived in Asia first, the most remarkable aspect of the sites in northern Israel is the evidence they contain of intentional burial. How can we be certain which skeletons came to rest in the caves and rock shelters through interment?

In establishing intentional burial, archaeologists look for relatively complete skeletons positioned as if they have been laid to rest, evidence that the skeletons have been deposited within pre-excavated pits and for associations with possible grave goods. The archaeologist Paul Pettitt has surveyed early evidence for burials around the world, and he argues that at least four early modern humans at Skhul and at least six at Qafzeh were buried. The skeletons are mostly complete, there is evidence for grave-cutting and, somewhat touchingly, both sites have a possible mother-and-child association.

For the Neanderthal sites, Pettitt argues that at least three individuals at Amud, one at Tabun and two at Kebara were buried. The female at Tabun may be associated with an infant, leading to speculation that the burial might have followed a death in childbirth. An unambiguous adult burial was excavated at Kebara by Bar-Yosef in 1982 (see plate 8). Dating up to 60,000 years old, it rested in a pit that was clearly dug into earlier archaeological layers, and the position of the body shows that it was deposited soon after death.

For most researchers this Neanderthal individual at Kebara settled an old debate over the question of whether the Neanderthals could speak. Despite the fact that the skull has not survived, the skeleton possesses a modern-looking hyoid bone, which is an essential part of the vocal architecture of *Homo sapiens*, and it could well have functioned in Neanderthals the same way as it does in us. With this hyoid bone, the Neanderthals may have been able to produce a range of sounds to form speech of great complexity.

Yet these burials do begin to hint at a difference in behaviour between the two varieties of human. This difference is seen not

It was once thought that only modern humans deliberately buried their dead, as seen here in a burial from Qafzeh, Israel (*right*), but scholars now believe that the Neanderthals did the same, as represented by examples from Kebara, Israel (*left*) and Shanidar, Iraq (*centre*).

only in the bones, but in what was buried with them. At Amud a Neanderthal child burial is associated with a deer bone, and at Kebara a Neanderthal child burial contains a rhinoceros tooth. As for the early modern human sites, a burial at Skhul is associated with a large wild boar mandible, and deer antlers were placed on top of a child burial at Qafzeh. The *Homo sapiens* were buried with slightly more prominent animal bones than the Neanderthals were, and this hints at a more important distinction.

At both Qafzeh and Skhul mollusc shell beads were found in possible association with the skeletons. Because the excavations took place a long time ago, without modern recording techniques, we cannot be sure that these were grave goods. We can be sure, however, that the people at Qafzeh and Skhul were using ornamental jewelry. Furthermore, red ochre was found in at least two of the graves at Qafzeh. Red ochre can be of ambiguous significance, because it can be used as an adhesive (e.g. to help bind a Levallois point to a wooden spear), as part of the preparation of hides and as a dye. Red ochre appears in European Neanderthal sites, such as Maastricht-Belvédère in the Netherlands from as early as 250,000 years ago. Archaeologists have argued that at Qafzeh it was used as a dye.

The use of jewelry and dye are both symbolic behaviours, which is one of the few behavioural differences we can discern, at least in degree, between early modern humans and Neanderthals some 100,000 years ago. Israel is not the only place one can find shell beads and red ochre. We can also find them at sites in Morocco, Algeria and South Africa, and this points to an African origin of the Qafzeh and Skhul people.

The humans at Qafzeh and Skhul were *Homo sapiens* like us, but they were not fully modern. Humans possessing a modern anatomy came out of Africa to Asia at a later date. The people of Qafzeh and Skhul were not the main ancestors of the modern population of the region. Instead we are left with the uncomfortable fact that this foray by early modern humans into Asia seems to have ended in failure. This upsets the common notion that the emergence of our species was a kind of inevitable march to human perfection.

Tens of thousands of years after Qafzeh and Skhul, Neanderthals lived in northern Israel and, as we discuss below, were pushing deeper and deeper into Asia. What happened to the *sapiens*? And where had they come from?

The great leap forward in Africa

For archaeologists looking for a point of origin for modern human behaviour and anatomy, it is confounding that the greatest collections of relevant archaeological sites are on the northern and southern coasts of Africa, separated by more than 8,000 kilometres (5,000 miles). This bi-coastal phenomenon is an intriguing one. It has not yet been established whether these behaviours first occurred simultaneously on the two ends of Africa or if modern humans spread them around the African coast. Archaeologists now think that multiple African populations may have contributed to the emergence of modern humans, rather than there being a single source area.

When modern humans finally made a successful transition from Africa to Asia, they spread quickly around the world, first along a coastal route all the way to Australia. It seems that coastal environments played a significant role in our species' emergence and growth. (This may be good news to nutritionists who posit a link between brain development and omega-3 fatty acids, which are found in fish.)

Before we look for the origins of fully modern humans, what can we say about the first wave of *Homo sapiens* to reach Asia? Back in the 1970s Stringer had linked the people of Qafzeh and Skhul to two African skulls, one from Omo Kibish in Ethiopia, the other from Jebel Irhoud in Morocco. Both were discovered in the 1960s and are now dated to 195,000 years ago (Omo Kibish) and 315,000 years ago (Jebel Irhoud). The palaeoanthropologists Katerina Harvati and Jean-Jacques Hublin have linked another Moroccan skull, this one from Dar Es-Soltan dating to 80,000 years ago, to this same group. Harvati has further argued that the types of teeth found in north Africa from this period closely resemble those at Qafzeh and Skhul.

From Morocco to Libya early modern humans developed a regional toolmaking tradition known as the Aterian. Aterian tools, derived from the African tradition of prepared-core technologies, are triangular and were probably mounted as spear-points.

Were the Qafzeh and Skhul people part of the Aterian culture? In the strict sense of following their stone tool manufacture, they were not. Their tools were Mousterian and were similar to those used by Neanderthals in the Middle East. Stone tools from a different site, however, do link inhabitants of the Middle East to Africa. In 2010 the archaeologist Jeffrey Rose discovered a cache of triangular stone tools in Oman from over 100,000 years ago which look surprisingly like tools from Sudan from the same period. While these tools, which were made by the Levallois technique, are not associated with human remains, many archaeologists view their close similarity as evidence of migration out of Africa. It is possible that in this period there were a series of small forays from Africa to the Middle East, and we are only seeing glimpses of it in the archaeological record.

Movement between north Africa and the Middle East was facilitated by the greening of the Sahara desert, one of the most significant effects of the Eemian interglacial. Enormous lakes and a network of rivers connected the north African coast with the rest of the continent, and the Sahara seems to have remained habitable until around 60,000 years ago. Archaeologists see the periodic opening and closing of the Sahara as a key part of how modern human populations were separated and re-connected several times during this dramatic phase of our evolutionary history.

North Africa has only recently emerged as an important region in human evolution. Its status in the human story comes not just from the bones, which are almost modern, but from some of the earliest evidence of a uniquely modern human behaviour. There are five sites in Morocco and Algeria that, like Qafzeh and Skhul, have produced mollusc shells with tell-tale puncture holes for use as jewelry. Similar shell beads have appeared as far away as Blombos Cave in South Africa.

A series of sites in South Africa, including Blombos Cave, Klasies River Mouth, Pinnacle Point and Sibudu Cave also contain early evidence of modern behaviour by *Homo sapiens*. In the early 2000s evidence for the earliest symbolism at Blombos Cave, on the southern coast of the continent, was scientifically dated to 100,000 years ago. This convinced most archaeologists that the roots of modernity, the so-called Upper Palaeolithic revolution that is well documented in Europe, can be traced to Africa. Blombos Cave contains evidence of a whole suite of modern behaviours, including ornamentation (with mollusc shell beads), fishing and a form of stone tool manufacture called pressure flaking (see the discussion of stone tools in Chapter 4), which the Neanderthals never used. Two of the most striking objects from this site contain abstract designs and are at least 70,000 years old. One is a plaque of red ochre which is scored, or engraved, with cross-hatches (see plate 10). The other is a stone fragment that is painted with red ochre. As we discussed in relation to Qafzeh and Skhul, red ochre can be used as an adhesive or as a dye. These objects at Blombos Cave make a strong case that its use here was symbolic, and that the early modern humans at the site were using paint to decorate objects and perhaps also their bodies. Archaeologists have also recovered what looks like a painting 'tool kit' and a processing workshop. Teeth recovered from Blombos Cave are almost typical for fully modern humans, sitting right on the edge of modern variation.

The cave sites of Klasies River Mouth from a similar date also contain human remains that are nearly modern. The Klasies bones show signs of possible cannibalism. Between Klasies River and Blombos is the site of Pinnacle Point, which has some of the world's oldest evidence of shell fishing (from more than 160,000 years ago) and also has red ochre.

The evidence from the northern and southern coasts of Africa points to a significant change among *Homo sapiens* which started before the Eemian and accelerated after it. Exploiting a coastal environment, humans were starting to use symbolism in the form of jewelry and red pigment over a wide area.

What is the significance of symbolism as a milestone in human development? In general, it is indicative of cognitive complexity – the ability to understand that a thing (e.g. a mollusc shell), in a certain context (e.g. a necklace), can have social meaning (e.g. the fact that the wearer has good taste and is well connected). The introduction of symbolism was probably also linked to more complex language, for example in the use of metaphor. Neanderthals at this stage were using red ochre, but it is not clear that this was a form of symbolism. (There is disputed evidence of earlier Neanderthal cave art – see p. 168.) Neanderthals may have started using bird talons for ornamentation during this period, but this was limited, with evidence from just a few sites. Ultimately symbolism shows an extension of social networks, which was a key competitive advantage modern humans had over groups that had not made the leap. Much later, Neanderthals also produced a cross-hatch pattern in a cave in Gibraltar (dubbed the first Neanderthal 'hashtag'), which, like the bird talons, may point to incipient symbolic thinking but was not nearly as extensive as the modern human use of shell jewelry.

At first modern humans seem to have made limited forays into Asia, reaching Israel and the Arabian Peninsula, but these were probably unsuccessful. Opinion differs on the exact date of the final, successful African exodus, but most researchers put it well after Qafzeh and Skhul. In 2015 researchers announced the discovery of a modern human skull from a cave in Israel near the Lebanese border dating to around 55,000 years ago, and this may be part of this wave of modern humans. At this time Eurasia was inhabited by Neanderthals, modern humans, Denisovans and possibly other groups such as *Homo floresiensis*. The stage was set for the final test – which human line was best adapted.

Meanwhile, back in Europe

As early *Homo sapiens* expanded from Africa into Asia around the time of the Eemian interglacial, what does the archaeological record say about developments in the Neanderthal homeland of Europe? The continent remained very much as it was before, with a population of Neanderthals using Mousterian stone tools. Their hunting prowess may have improved, and there is evidence that they could take down an impressive array of animals: bison, bear, mammoth and the exceptionally large two-horned Merck's rhinoceros. There is also some evidence they ate other Neanderthals, although there has been much controversy about this.

While there are many European sites from this time with Mousterian stone tools, there are only a few collections of human remains that are securely dated to the Eemian and the time immediately after it. Most notable is a cache of more than 800 human remains excavated from Krapina, a collapsed cave in Croatia close to the border with Slovenia, from 1899 to 1905. A site in France, Moula-Guercy, which contains several dozen Neanderthal bone fragments, has also been dated to this time.

At the time of its discovery Krapina represented one of the greatest collections of ancient human bones ever found, and it certainly retains its place in that elite class, comparable to Sima de los Huesos at Atapuerca. Estimates of the number of individuals represented at Krapina have gone as high as seventy-five or eighty, but have since been revised downwards to about one third of that total. The bones are gracile for Neanderthals and highly fragmented. Some bones show signs of osteoarthritis and healed fractures.

As is the case at Sima de los Huesos, there have been various theories about how such a disarticulated mass of human bones found its way into Krapina Cave. Due to cut marks evident on many of the bones, the excavator, Dragutin Gorjanović-Kramberger, suggested that they had been cannibalized. More recently, others such as Eric Trinkaus and Mary Russell have proposed that the cut marks came about through ritualized defleshing and secondary burial of the dead rather than cannibalism. The anthropologist David Frayer, meanwhile, takes the middle ground, arguing that most of the cut marks are evidence of cannibalism while some indicate ritual.

As we argued in Chapter 2 when we discussed cases of cannibalism at Gran Dolina, we believe that cannibalism is the most logical explanation for the cut marks on these bones. Neanderthals who were used to butchering large game would have had little problem in butchering other Neanderthals. We know from studies of Neanderthal teeth that episodes of severe starvation were not uncommon, and it would be shocking if they let protein go to waste, especially after taking the trouble of defleshing their deceased kin.

The archaeologist Timothy Taylor has noted in his book *The Artificial Ape* (2010) that despite widespread evidence for cannibalism in prehistory, including cannibalism by fully modern humans, researchers often dismiss strong ethnographic, historical and archaeological evidence for the practice, possibly from a well-meaning anti-racism. Trinkaus, for example, called cannibalism 'that dreaded, bestial practice', and argued that archaeologists who wished to show a clear separation between us and them employed the interpretation as a sort of smear against the Neanderthals. Cannibalism may be 'an ugly spectre', in Trinkaus's terms, but it requires a great deal of mental gymnastics to believe that extensive defleshing went on at Krapina without any subsequent consumption of that flesh. And we must remember that the practice of cannibalism may separate our prehistoric forebears from us today, but it certainly does not represent a difference between Neanderthals and modern humans. Human flesh can be eaten for a wide range of reasons, such as respect for the dead or as part of the bounty of a victory in conflict. The practice does not necessarily mean that Neanderthals lacked respect for their own kind. What it probably does reflect is that they went through difficult periods of protein shortage.

Debate has continued over the question of cannibalism at Krapina. Our own interpretation has been given some scientific legitimacy thanks to a re-examination of the bones by the anthropologists Tim White and Nicholas Toth. White and Toth established formal criteria for cannibalism based on the cut marks and deposition of the bones and on the preferences for which bones were selected for cutting, and they compared patterns of Neanderthal butchery of game animals with what can be observed on Neanderthal bones. From a series of such studies they argue that Krapina is one of many Neanderthal sites in Croatia, France and elsewhere that show significant evidence for cannibalism.

White was also part of a project in the 1990s at Moula-Guercy, a cave site above the River Rhône in France, where evidence for cannibalism has been more widely accepted. The butchered remains of six Neanderthal individuals at Moula-Guercy have been dated to more than 100,000 years ago. The processing of the Moula-Guercy bones was a little more thorough than at Krapina.

Krapina has historically been as much a mirror of our own prejudices as it has been a key Middle Palaeolithic site. Aside from the cannibalism debate, there is the fact that it was long neglected by an archaeological community more focused on sites in France and Germany than on the European periphery. Of greater importance for our purposes was the key part that Krapina played in addressing the question of a possible Neanderthal role in the evolution of modern humans.

In the early 20th century, before radiometric dating techniques were available, Harvard anthropologist Earnest Hooton suggested the label 'classic Neanderthals' to describe the more robust specimens from western Europe that we will meet in Chapter 6. He thought that the lighter-boned Neanderthals of Krapina represented a different race. We now know that the Krapina Neanderthals pre-dated the 'classic Neanderthals' by tens of thousands of years. Hooton's work on human variability relied unduly on the concept of race, in perhaps one of the less proud phases in the history of anthropology, and it is no surprise that he interpreted human evolution in racial terms.

Fast-forward to the 1970s, when Milford Wolpoff studied the teeth from Krapina and argued that they were on the evolutionary path towards modern human teeth. In subsequent years Wolpoff developed the multiregionalist theory of human evolution, which argued that since the days of *Homo erectus* humans around the globe exchanged genes within a single breeding population, despite the differences in appearance that others attributed to a proliferation of different species. In short, Krapina helped convince Wolpoff that there is a little Neanderthal in all of us.

The countervailing narrative to this view was put forward by Chris Stringer, who also studied the Krapina material in the 1970s. Under the replacement theory, which is also called Out of Africa or Recent African Origins (Stringer's preferred term), he proposed that *Homo sapiens* evolved in Africa and replaced all other varieties of human, including Neanderthals. Stringer's theory is supported by the fact that European Neanderthals were evolving towards the 'classic' type and away from any similarity with modern humans.

The fate of the Neanderthals takes centre stage in Chapter 6. Were they among our own direct ancestors or did *Homo sapiens* replace them and cause their extinction? Thanks to modern DNA technology, we may finally have an answer to this question. Curiously the answer has enabled both Stringer and Wolpoff – supporters of Out of Africa and the multiregionalists respectively – to claim vindication.

In 2015, Krapina hit the headlines again, when eagle talons from the site were found to have wear patterns consistent with use as jewelry. Perhaps these Neanderthals were starting along the path of the modern humans in Africa with their shell ornaments. This isolated case shows that Neanderthals may at least have had the potential for modern behaviours.

LAND OF RED CLAY

Archaeologists love to tell stories about how they stumbled upon significant discoveries. These stories often involve stopping the car for reasons that have little to do with spotting an unexplored new site out of the window. The site of Kokkinopilos in Epirus in north-western Greece has a special place in this genre of storytelling. At the moment of its serendipitous discovery, few could have predicted the site's importance both as the largest Neanderthal site in the region and as a training ground for young archaeologists over several generations.

The year was 1962 and Cambridge prehistorian Eric Higgs had brought a group of students to Greece in his infamous Land Rover in the hope of finding some new Palaeolithic sites. As a project director, Higgs could become so absorbed in his work that he rarely thought of practical matters such as feeding his crew. His students learned to scrounge whatever food they could find, running into shops at any opportunity and hoping Higgs would not drive away without them. Part of the reason Higgs chose Greece was that it did not present as many difficulties as Libya, where a Cambridge team had been working for years under Charles McBurney at the Haua Fteah cave near Benghazi.

Higgs explored high into the Pindos Mountains in Epirus. In the town of Metsovo, home to ethnic Vlachs who speak a language similar to Romanian, one of Higgs's students, Rhys Jones (who went on to have a distinguished career in Australia), managed to communicate with the locals by drawing on his knowledge of Latin, which they seemed to understand. Deep in prehistory Neanderthals did not make it to such a high elevation in the mountains, and the team found little of interest there.

Higgs drove the group towards the coastal lowlands, and as they came down to the foothills via the main road from Ioannina, the picturesque regional capital in the mountains, one of the team members identified what he thought was a cave, and Higgs stopped the Land Rover. The students no doubt hoped that this cave might hold riches similar to Haua Fteah, which had deposits from classical times all the way back to 200,000 years ago.

But instead – a sad indictment of their observational skills – what Higgs's students found in Epirus was no cave. It was a tunnel, dug into the living rock in Roman times as part of an aqueduct which provided the nearby city of Nikopolis with fresh water. In one version of what happened next, Higgs sent Jones to the top of the hill above the tunnel to see what might be of interest there. In another version, Jones simply needed a moment of privacy.

What Jones saw was unexpected in this lush area of goat herders and small farmers. He found a vast moon-like badland with thick reddish-yellow sandy and clay deposits, eroding into deep gullies which exposed a wealth of textbook-quality Mousterian artifacts. The team had finally found what it was looking for in Greece.

The group returned to Ioannina to inform Sotirios Dakaris, the head of the archaeological service for the region. According to Jones, Higgs had kept the artifacts they had initially found, but hid them in the car's spare tyre, because they were on a survey mission and had not been given permission to collect anything. When Dakaris later visited the redbeds with the team, a nod from Higgs told Jones to surreptitiously grab the ancient tools and 'salt' the site with them to demonstrate the great potential of further investigations there.

Dakaris was thrilled, and he and Higgs dug two test trenches and collected further samples. These stone tools formed the basis for the first professional publication of Paul Mellars, who later became professor of archaeology at Cambridge, a world-leading expert on Mousterian tools and author of *The Neanderthal Legacy* (1996).

In subsequent decades the site of Kokkinopilos (literally 'red clay' in Greek) and other, smaller ones like it on similar redbed badlands in the area formed the basis for repeated research visits by some of the big names in Palaeolithic archaeology, including Higgs, Mellars, Geoff Bailey, John Gowlett, Clive Gamble, Curtis Runnels and Tjeerd van Andel.

As well as providing research fodder for such eminent prehistorians, Kokkinopilos's impact also extended into popular culture. The novelist Hammond Innes used it as a key setting in the thriller *Levkas Man* (1971), which centres around a controversial archaeological theory by Pieter Van Der Voort, a character inspired by Higgs. In the dramatic conclusion Van Der Voort is found hiding at Kokkinopilos, where he had been using the deep eroding gullies as cover.

The unromantic reality of Kokkinopilos and most of the sites in the area is that stone tools comprise the vast bulk of surviving artifacts. But in an eerie anticipation of the

A concrete hut overlooking the Louros River Valley from the site of Kokkinopilos, Greece, where numerous Mousterian stone tools have been discovered.

discovery of Cosquer Cave near Marseille, France, in 1985, Innes imagined in the book that early paintings were preserved on the nearby island of Levkas in a cave accessible only by divers. The archaeology of Epirus and the Ionian Islands may not be quite that exciting, but it has played a key role in our understanding of Neanderthals and modern humans in the region and in south-eastern Europe more generally.

Why are so many Neanderthal tools exposed by the erosion of red soil deposits in north-western Greece? Most likely the redbeds are residual deposits from ancient lakes or seasonal ponds which acted as traps for sediments washed down by erosion from the surrounding mountain slopes. Millennia of earthquakes and tectonic uplift have long since drained the water and left behind the sterile red clay. The stone tools are evidence of Neanderthals visiting the lakes to catch prey that would have been attracted to the water.

Kokkinopilos is the mother of these sites – the first one to be discovered, the biggest one and the one with the widest range of tools. At other, smaller sites in the region, Neanderthals used a variety of techniques to manufacture Mousterian tools. Kokkinopilos is the only site that contains evidence of all these techniques. This means that the site was probably teeming with wildlife and attracted Neanderthals over many millennia.

In addition to the Roman aqueduct that runs underneath Kokkinopilos, one curious modern architectural feature on the site is a small concrete hut which acts as a shelter for shepherds. According to the locals this structure, which overlooks the road, was used by German soldiers during the Second World War to guard the main supply route to Ioannina. The Nazis, it seems, used the site in much the same way as the Neanderthals – as the ideal spot to observe movement through the narrow ravine that is the main route from the coastal plains into the mountains. Neanderthals, of course, would have had access to the ancient lake which has long since disappeared, and they would have been tracking animal herds in the valley, while the Nazis were looking for resistance fighters.

Pushing east

We began this chapter with the image of a Neanderthal at Gorham's Cave, Gibraltar, looking out across the water and not quite being able to see modern humans inhabiting the north coast of Africa. Modern humans using symbolic behaviours had exploited a coastal environment and expanded along a north–south axis, stretching from Morocco to South Africa. How far did the Neanderthals go as they pushed eastwards during this period?

Neanderthals were expanding their range to around double what it had been previously. In addition, there is evidence they were burying their dead much as modern humans were. And their brains continued to grow larger, with an individual from Amud dated to 45,000 years ago carrying over 1740 ml (60 oz) of brain, which is at the extreme upper end of human variation.

Tabun, Kebara and Amud in northern Israel are some 4,000 kilometres (2,500 miles) east of Gibraltar. About 1,000 kilometres (600 miles) further east we arrive at the cave of Shanidar in Iraqi Kurdistan, close to the Iranian border, where nine Neanderthal skeletons were unearthed in the 1950s by Ralph Solecki of Columbia University (evidence for a tenth and eleventh individual, and possibly even a twelfth, was discovered in subsequent studies). You have to travel the same distance again to reach Teshik-Tash, Uzbekistan, where a Neanderthal child was excavated by the Soviet archaeologist

Alexey Okladnikov in 1938. Another 2,000 kilometres (1,200 miles) further east still, thanks to DNA testing on bones of ambiguous affiliation from Okladnikov Cave (named after the Siberian-born excavator of Teshik-Tash), and bones from nearby Denisova Cave (which some say was named after a hermit called Denis), we now know that Neanderthals who were closely related to their European kin reached the Altai Republic of Siberia.

It is ironic that as the Neanderthals headed towards extinction, they were also extending to their maximum range, and doing so in inhospitable areas such as Siberia while the global climate was trending colder. Shanidar is dated to 70,000–40,000 years ago, Teshik-Tash to the earlier end of this range and Okladnikov to the later end. The Neanderthals ultimately reached about the same distance of 8,000 kilometres (5,000 miles) along an east–west axis that *Homo sapiens* achieved on a north–south axis within the confines of Africa. As we will see in the next chapter, by the time of the later Neanderthal dates in Siberia, modern humans had not only left Africa but had worked their way through the Neanderthal homeland of Europe.

The humanity of the Neanderthals was becoming increasingly evident in these Asian sites. Citing discoveries at Shanidar and Teshik-Tash, archaeologists started attributing to the Neanderthals human qualities such as compassion and sentimentality. Not all of the evidence has stood the test of time, but there remains

Shanidar I, the skull of a male Neanderthal found in the cave of Shanidar, Iraq. This individual had suffered substantial injuries, which had partially healed, suggesting that he had been cared for by others in the group. He was the inspiration for Creb, the disabled shaman in Jean M. Auel's novel *The Clan of the Cave Bear* (1980).

Shanidar Cave in Iraq. At least eleven Neanderthal individuals have been discovered in this cave, dating to 70,000 to 40,000 years ago.

plenty of support for the notion that Neanderthals shared many of the qualities that we once thought separated us from them.

At Shanidar the remains of at least eleven individuals (some buried, others apparently crushed by rock falls) were found in association with hearths, indicating that they lived, died and were sometimes buried in this location. One of these individuals, labelled Shanidar I, was severely disabled in life, judging by partially healed wounds that most likely came about through trauma. The fact that this individual survived debilitating blows to his arms, legs and head shows that he benefited from intensive support.

Shanidar gained its initial fame from another individual, Shanidar IV. The excavators argued that Shanidar IV was intentionally buried and that the grave was adorned with a great heap of flowers. We discuss this burial again in Chapter 7 for its role in inspiring a particular literary image of the Neanderthals. The flower burial theory has since been questioned, not just for the sheer implausibility of cut flowers being preserved in the archaeological record, but thanks to a better understanding of how pollen is preserved in ancient soil levels. In any case Shanidar I provides compelling evidence that the Neanderthals were a caring form of humanity.

Starting in 2014 a team led by Graeme Barker, Christopher Hunt and Tim Reynolds became the first to excavate at Shanidar since the 1950s. In 2020, they published evidence of the eleventh Neanderthal at the site and hope to shed further light on the question of intentional burials. They also indicated that there may be other remains yet to be unearthed in the cave. After the terrorist group ISIS was pushed out of the area, Shanidar has become popular with visitors.

Sources of meat around Shanidar would have included wild sheep and goat which were native to this area. A site just outside the cave provides some of the world's earliest evidence of domestication of these animals, which occurred during the modern human occupation of the Middle East in the Holocene. For modern humans, Shanidar is an important site in the development of civilization, while for Neanderthals it appears to have been an isolated outpost.

At Teshik-Tash a child of no more than ten years of age was found in association with ibex horns in what first appeared to be one of the clearest Neanderthal ritual burials. Archaeologists now doubt whether this was a deliberate burial, noting that the child and the horns might have been dragged into the cave by hyenas. Taken together with the Neanderthal remains at Okladnikov and Denisova Caves, these two sites

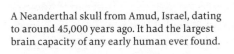

A Neanderthal skull from Amud, Israel, dating to around 45,000 years ago. It had the largest brain capacity of any early human ever found.

show that the Neanderthals could penetrate inhospitable conditions in the mountains of central Asia.

Whether or not the child of Teshik-Tash was buried, it is fairly clear that the Neanderthal skeleton from Kebara, Israel, 60,000 years ago, was a deliberate burial. Coincidentally, this was within the time range that geneticists place the exodus out of Africa that brought about the *Homo sapiens* domination of the planet.

As the Neanderthals reached the peak of their achievements, they came in contact with modern humans leaving Africa. But there was also another group in Asia, one we did not know about until the invention of technology to read ancient DNA. Before we meet this third human population in Asia from this time, the Denisovans (and there may have been others confined to islands in Indonesia), let us first look at the key breakthroughs in DNA.

Cracking the DNA code

As we have seen, Neanderthal bones contain an enormous library of information. Their shape and thickness can tell us about Neanderthal bodies and also hint at evolutionary connections with other hominin species. The ratio of the unstable carbon isotope ^{14}C to the stable isotope ^{12}C can tell us how long ago the individual died. And the ratio of stable carbon and nitrogen isotopes can tell us what the individual ate.

There is one more piece of information contained within some Neanderthal bones, and this is the hardest to identify among possible contaminants, for it is a structure common to all living things. DNA is the building block of life: it contains the genes that define an individual or a species. Some Neanderthal bones still carry Neanderthal DNA. Accessing it, however, is far from easy. The people who excavate and analyse Neanderthal bones have DNA that is extremely similar. Bacteria that live in the bones also have DNA.

Ironically, the pioneering researcher who first extracted Neanderthal DNA only did so after becoming frustrated with the problem of contamination in extracting modern human DNA from more recent remains. Plus, 'Neanderthal DNA seemed like the coolest thing imaginable to me', wrote Svante Pääbo in *Neanderthal Man: In Search of Lost Genomes* (2014), his account of his great breakthroughs in palaeogenomics. At the Max Planck Institute for Evolutionary Anthropology in Leipzig, Germany, Pääbo and his team have found ways around the challenges of contamination to produce a draft, and subsequently a high-resolution Neanderthal genome.

The first clue about Neanderthal DNA was published in 1997 in the periodical *Cell*. Pääbo and his colleagues were able to recover mitochondrial DNA (mtDNA) from the original bones found in the Neander Valley back in 1856. Mitochondrial DNA is easier to recover and sequence because it is much smaller than nuclear DNA. The drawback of mtDNA is that it can only look back along the maternal line, much like tracing genealogy only through last names which are passed down the paternal line, thus excluding information from other direct ancestors.

The first two contributions of mtDNA to the Neanderthal story were to shed light on the question of possible interbreeding and the question of the time of our common ancestor. In the two decades following the initial publication, conclusions springing from both of these studies were overturned. This is a reminder that new scientific technology often takes years before its results become robust enough to rewrite the textbooks.

The first, most striking impact of the original publication of Neanderthal mtDNA was that it seemed to confirm the theory that modern humans coming out of Africa completely replaced all archaic species, including the Neanderthals. Compared to modern human mtDNA – our mtDNA today – the Neanderthal version looked very different. This means that Neanderthal women do not have any direct female-line descendants (i.e. mother to daughter to daughter to daughter, etc.) living today.

Many researchers concluded that this result implied that no Neanderthals have any living descendants. But one can easily see the problem with this kind of thinking. Mother-to-daughter chains can break fairly easily in any line of descendants.

The other contribution of the first mtDNA study was to estimate modern humans' evolutionary distance from Neanderthals. Because mtDNA is passed through the maternal line, unmixed with contributions from the father, and is believed to mutate at a constant rate, it is possible to calculate the time since two individuals shared a common ancestor. By the early 2000s, the genealogical potential of mtDNA was promoted and popularized by the Oxford geneticist Bryan Sykes in his book *The Seven Daughters of Eve* (2001). He grouped modern Europeans into seven groups of descendants (haplogroups), distinguished by their mtDNA signatures. Though distinguishable, these seven groups are closely related to each other. People of European descent who purchased an early genetic test could trace their maternal ancestry back a few thousand years through this technology.

Researchers used similar technology to trace mtDNA back to a founder population of modern humans that had left Africa at around the same time, some 80,000–65,000 years ago. Non-African mtDNA has such low variability that it seems that humans living today with ancestry from Europe, Asia, Australia and the Americas can all trace their maternal lines to this small group. Looking further back in Africa is the so-called Mitochondrial Eve, the maternal ancestor of all modern humans, estimated to have lived around 150,000 years ago.

By the later 2000s, researchers looked to extend this result even deeper into the past, to the last common maternal ancestor with Neanderthals. In 2008 headlines heralded some key results arising from the study of Neanderthal mtDNA recovered from eleven Neanderthal individuals from the Neander Valley (Germany), Mezmaiskaya Cave (Russia), Vindija (Croatia), Engis and Scladina (Belgium), La Chapelle-aux-Saints and Rochers de Villeneuve (France), Monti Lessini (Italy) and El Sidrón (Spain). According to this line of evidence, Neanderthals and modern humans last shared a common ancestor around 500,000 years ago. (This number has since been pushed back to closer to 700,000 years ago.) This seemed to reinforce the fossil evidence, which indicated that the African and European lines of *Homo* had shared a recent ancestor around the time *Homo heidelbergensis* appeared in Europe. It also confirmed Pääbo's earlier result that there was no subsequent interbreeding between the species.

Just as these results came out, the picture began to change dramatically. Thanks to two technological developments, the recovery of Neanderthal DNA was proceeding apace. Polymerase chain reaction (PCR) was developed in the 1980s as a way to amplify the signal of a limited amount of DNA and make possible the analysis of small, degraded fragments from ancient bones. And high-throughput sequencing was developed by the company 454 Life Sciences, founded in 1999 in Branford, Connecticut, making it economical in terms of time and money to produce a draft genome, comprised of nuclear DNA rather than the much smaller mtDNA.

A draft Neanderthal genome, based on nuclear DNA, was indeed published in 2010 in the periodical *Science*, and with it came a shock announcement. Based on samples extracted from three bones from three different individuals from Vindija Cave in Croatia, Pääbo and his team argued that there had been interbreeding between the two populations. They went further and said that modern human populations not indigenous to Africa (from modern Europeans to Aboriginal Australians) show a consistent 1–4 per cent intrusive Neanderthal DNA in their genomes (this figure has since fallen closer to 2 per cent), while Neanderthals did not show intrusive DNA from *Homo sapiens*. In addition, they argued that modern humans indigenous to Africa do not have Neanderthal DNA at all. (Subsequent studies in 2020 showed that Africans, and therefore all humans living today, carry some Neanderthal DNA, although the African population has a much smaller percentage, representing possible back-migration.)

These results point to at least one interbreeding event that took place at the early stages of the out-of-Africa expansion, before modern humans reached Europe or Australia. It did not take long for this discovery to be commercialized. A number of companies soon offered a Neanderthal DNA percentage test as a standard part of their DNA testing service.

A more detailed picture of the interbreeding between Neanderthals and modern humans emerged with the sequencing of a high-resolution genome from the femur of a modern human individual who lived in Siberia 45,000 years ago. This was the first case in which DNA has been retrieved from an early modern human – itself a breakthrough in ancient DNA studies. This modern human does not have more Neanderthal DNA than present-day humans, but the Neanderthal DNA that he has is preserved in longer chunks. This allowed researchers to pinpoint the timing of the interbreeding to between 50,000 and 60,000 years ago. There may have been more recent interbreeding events between the two populations, but their effect was less significant. Equally, potential interbreeding events between Neanderthals and early modern humans who lived in the Middle East around 100,000 years ago do not seem to have made any significant impression on the genome of present-day humans. A partially preserved modern human skull found in Manot Cave in northern Israel, dating from around 55,000 years ago, may be representative of the modern human populations that interbred with Neanderthals after the dispersal out of Africa and before expanding across Eurasia.

The discovery in 2010 of Neanderthal nuclear DNA and inter-breeding with modern humans was not the only shocking announcement from that year. While the Neanderthal genome was published in May, 2010, December brought the news that there was another group, called the Denisovans after a cave in Siberia that contained a finger bone and tooth with genetic signatures different from both modern humans and Neanderthals. From the mtDNA of Denisovans, there seemed to be a link with the incipient Neanderthals of Sima de los Huesos (see pp. 59–63), suggesting a recent common ancestry between Denisovans and Neanderthals. Furthermore, Denisovan DNA was identified in the genes of present-day Melanesians, comprising up to 6 per cent of their genomes.

We are left with a complex picture in Asia, with the intermixing of at least three human populations that had been previously separated for perhaps hundreds of thousands of years. The next chapter will look at how this process may have unfolded, leading to the survival of only one human species. But first, we will explore the growing body of evidence of the mysterious Denisovans, whose existence was not even suspected when we started writing the first edition of this book.

The Denisovans

The Denisovans have the distinction of being the first human variant, or perhaps separate species of *Homo*, that was first identified not from the particular shape of their bones and bodies, but from their DNA. In just a decade, evidence for them has grown significantly from the little finger from Denisova Cave that started it all. Yet there is still far less evidence for them than we had for the Neanderthals in the mid-19th century.

Denisova Cave, in the Altai Mountains of Siberia, Russia, contains a Middle Palaeolithic occupation that dates back to around 300,000 years ago. It is where the Denisovans were first discovered and remains the site that has produced most of the known evidence. The discovery of relatively large amounts of Denisovan DNA in modern southeast Asian populations, however, suggests that they were once more widespread across Asia. At this stage it is not possible to say whether the Denisovans were a sort of eastern Neanderthal, an Asian descendant of *Homo heidelbergensis*, a relative of 'Galilee Man' (neither Neanderthal nor modern) from Zuttiyeh Cave in Israel or something else we have not considered. What is clear is that there were at least four kinds of human in Asia at this time – Neanderthals expanding eastwards, *Homo sapiens* coming out of Africa, Denisovans in Siberia and perhaps other places and *Homo floresiensis* in Indonesia.

The first evidence of Denisovans came in 2010, when DNA was published from part of a finger bone. The bone itself had been excavated in 2008, along with molars that appeared to be large and archaic in shape. There is something of an archaeological mystery surrounding part of the finger bone. It was discovered by Russian archaeologist Anatoly Derevianko, who sent parts of the bone to two different labs. The Denisovan DNA was discovered and published by Pääbo from the piece sent to the Max Planck Institute in Leipzig. The other part of the finger went to the Lawrence Berkeley National Laboratory and subsequently to two other labs, and it disappeared somewhere in its travels. Nevertheless, a reconstruction of the full finger fragment, before it was split up and studied, shows a surprisingly thin, modern digit.

In 2016, a subsequent excavation at Denisova Cave uncovered two pieces of a Denisovan braincase. Aside from appearing robust and archaic, like the molars found earlier, the bones were not complete or distinctive enough for researchers to say much about the Denisovans' place in the human family.

Denisova Cave was an attractive location for at least three different human populations: Neanderthals, Denisovans, and modern

8. The skeleton of a Neanderthal found in Kebara Cave, Israel. Dating to around 60,000 years ago, it provides key evidence that Neanderthals sometimes buried their dead.

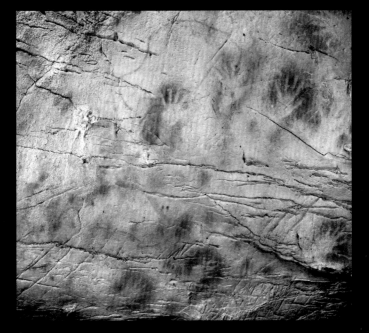

Opposite
9. Reconstruction of a Neanderthal butchering a deer, based on remains found at Shanidar Cave, Iraq.

Top and Above
10. Examples of early symbolic behaviour: a red ochre 'plaque' with cross-hatches from Blombos Cave, South Africa, dating to around 100,000 years ago, and hand stencils at El Castillo, Spain, dating to more than 40,000 years ago.

Opposite
11. Reconstruction of a Neanderthal based on the La Ferrassie 1 fossil. Reconstruction by Elisabeth Daynès.

Above
12. Neanderthal skull and long bones from Saint-Césaire, France. The discovery of this 'classic' Neanderthal dating to around 40,000 years ago helped to demonstrate that modern humans and Neanderthals represented separately evolving lines rather than sequential stages in our own evolution.

Right
13. Neanderthal skull and parts of a skeleton found in 1908 at La Chapelle-aux-Saints, France. The individual suffered from arthritis and bone degeneration, which led to early reconstructions of Neanderthals with a stooping posture.

Opposite
14. Archaeologists excavating in El Sidrón Cave, Spain, which was discovered in 1994 and contained the remains of thirteen Neanderthals, from which Neanderthal DNA has been extracted.

Left
15. Svante Pääbo of the Max Planck Institute, Leipzig, Germany, a pioneer in the field of palaeogenetics and the driving force behind the discovery of Neanderthal DNA.

Below
16. Antonio Rosas at El Sidrón Cave, where he leads the analysis of the most complete collection of Neanderthal remains yet found in Spain.

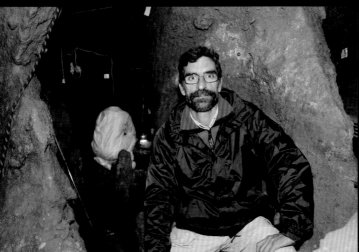

17. The Neanderthal Museum in Germany used prosthetics and make-up for this 'Stone Age Clooney' in order to re-examine the question of whether Neanderthals, if they dressed like us, would be able to walk among us without undue attention.

18. A group of Neanderthals from the film *The Clan of the Cave Bear*, made in 1986 and based on the bestselling novel by Jean M. Auel.

humans. Because many of the human remains – over a dozen – are in tiny fragments, it is thought that they represent the remains of a hyena den, so not all of the human inhabitants of the cave arrived there voluntarily. The fragmentary nature of so many of the bones in Denisova Cave has been a challenge for archaeologists. A technique called zooarchaeology by mass spectrometry (ZooMS) enables researchers to identify collagen that is likely to come from the human line, as opposed to other animals. Then the ancient DNA experts come in, knowing which tiny fragments hold the most promise for further study.

Bones from the cave have not only provided the first DNA evidence for Denisovans, but they have also given us the second Neanderthal genome. Perhaps the most exciting DNA evidence from the cave came in 2018 and involved both Neanderthals and Denisovans. Nicknamed Denny by Svante Pääbo, Viviane Slon and their team, a female from around 90,000 years ago (whose bone fragment was first flagged for study by ZooMS) seems to have had a Neanderthal mother and a Denisovan father. This is the most dramatic evidence yet found for interbreeding between human populations.

Evidence of the Denisovans is emerging in other places. In 1980, a Buddhist monk went into Baishiya Karst Cave on the Tibetan Plateau in order to meditate. While there, he found a jawbone, including two unusually large molars. For decades, the jawbone remained a curiosity stored at Lanzhou University in China. Then, in 2019, researchers dated it to 160,000 years ago and identified it as Denisovan. No DNA was extracted from the ancient bone, but scientists used a technique called proteomic mass spectrometry (similar to ZooMS, above) to examine the collagen protein in the teeth, and this had a Denisovan signature. Subsequent excavations at the site turned up more recent Denisovan mtDNA, from sediments rather than fossils, linked to the population in Siberia. Thus, Tibet became the second location in Asia with identified Denisovan remains.

The Tibetan monk and his Denisovan predecessor in that cave may have had another thing in common. Tibetans today are adapted to life at high altitudes with the help of a gene that regulates haemoglobin. It turns out that Denisovans had the same gene, and modern Tibetans likely inherited it from them through interbreeding between the two species. It is amazing to think that Denisovans adapted to the Tibetan plateau and then enabled modern populations to shortcut the evolutionary process by simply inheriting the gene from the human group that got there first.

An additional intriguing fossil emerged in 2018 in north-eastern China. In an astonishing story of resilience and memory that begins in 1933, a Chinese labourer under Japanese occupation discovered an ancient skull. He hid the skull at the bottom of a well and waited until 2018, when he was nearing death, to tell anyone where to look. Researchers from the Geoscience Museum in Hebei GEO University, China, where the skull is now held, date it to more than 146,000 years ago and call it *Homo longi*, or 'Dragon Man'. Everyone agrees that this skull is neither modern nor Neanderthal. Many researchers outside of China think it may be the first full Denisovan cranium, giving us evidence of how they might have looked.

Aside from the bones, the balance of evidence we have for Denisovans is in modern DNA. In contrast to our Neanderthal inheritance, which peaks at around 2 per cent, the Denisovan inheritance goes as high as 5–6 per cent. Curiously, the modern populations with the highest percentage of Denisovan DNA are in Melanesia, and also among indigenous peoples of the Philippines.

A study in 2019, led by computational biologist Murray Cox, found that there are three separate lines of Denisovan DNA in modern populations. The team found two distinct lines among modern people in Papua New Guinea, and a third line in East Asia. This result indicates that while Asia was the focus of a great deal of interbreeding (Neanderthals and Denisovans, Neanderthals and modern humans, modern humans and Denisovans), it is also a continent large enough that it has historically divided populations for many millennia.

The most intriguing discovery that may be linked to the Denisovans is a cross-hatch pattern engraved on a bone, with traces of red ochre, found in Henan Province, China, and dated to more than 100,000 years ago. At this point the link with the Denisovans is purely circumstantial, as it could have been made by Neanderthals or by that early wave of modern humans leaving Africa. The cross-hatch pattern is reminiscent of an artifact from South Africa and cave art in Spain (see plate 10). This Chinese artifact is older, however, and reminds us that we know so little about Asia and the Denisovans that there are surprises yet to come.

For the three major lines of human species at this time, Asia was a global crossroads. The first opportunity for the Neanderthals to meet *Homo sapiens* came in Asia, when early modern humans first left Africa and Neanderthals spread eastwards. This early African exodus by *Homo sapiens* was apparently not successful, and as global temperatures returned to cold, glacial conditions, the Neanderthals had greater success in advancing across the world's biggest continent. In Asia, the Neanderthals met the Denisovans, who seem to have shared a common ancestor with them more recently than the Neanderthals shared one with *Homo sapiens*.

Having survived glacial cycles in Eurasia since around 600,000 years ago, the Neanderthal line was about to enter its final stage. This coincided with the arrival of the next wave of *Homo sapiens*. In Chapter 6 we attempt to answer the question why events this time unfolded quite differently from the first Asian arrival. We will also take a close look at the classic Neanderthals and their ways of life, for it is in their final stage of existence that the Neanderthals became most distinct from their African cousins and took on the form that most people associate with this fascinating human species.

Endgame

60,000 to 25,000 years ago

If the Neanderthal story were a movie, this would be the dramatic final act. It had incredible twists and turns – wild swings in temperature, extinctions of large European mammals and the intrusion of a rival human species into their homeland. It is the period for which we have by far the most evidence of their way of life, and yet it is a time that remains full of mystery.

This period falls into three distinct phases. The phase of the 'classic Neanderthals' began 60,000 years ago when the Earth's climate entered a mild yet highly variable interglacial, which had nothing like the kind of warmth and stability that characterized the Eemian or the current Holocene. The Neanderthals took advantage of these conditions to increase their range within Europe, where they repopulated the northern areas, and they continued to spread eastwards into Asia.

The second phase started some 45,000 years ago when our species arrived in Europe. This phase, which has long been the subject of both literary and archaeological speculation, has in recent years come into better focus. A key problem with this period is that it takes place right at the horizon of effective carbon dating. A number of recent scientific advances – improvements in dating technology, the generation of fine-resolution data of past climate and the introduction of computer modelling – have shed new light on one of the most pivotal moments in human evolution.

Perhaps the greatest surprise to come out of this new research is that modern humans did not initially have much of an advantage over the Neanderthals.

The final phase starts around 37,000 years ago. At this time the modern human presence in Europe underwent a dramatic change. An archaeological culture called the Gravettian, probably representing an influx of new people, appeared, and was more successful than both the Neanderthals and the earlier wave of *Homo sapiens* that had been in Europe for nearly 10,000 years.

By the time of the spread of the Gravettian culture through Europe, the Neanderthals were extinct. After hundreds of thousands of years as the masters of Europe, our human cousins were gone. All they left behind were the less perishable traces of their existence, their stone tools, hearths, food debris and of course their bones, which we would later use to date their extinction and to extract their DNA.

To put these events in perspective, it is helpful to imagine the entire course of human evolution, from the appearance of the first *Homo habilis* in Africa to the present, as taking place over the course of a single day. For convenience, let us start the clock with midnight representing 2.4 million years ago, which is within the range of when the genus *Homo* is thought to have emerged. In this time-compressed day, with each hour representing 100,000 years, humans left Africa at dawn, around 5 to 6 am. They first arrived in Europe at noon, already at the halfway point since the appearance of the human line. In what was probably a subsequent out-of-Africa expansion, *Homo heidelbergensis* or its ancestor species spread from Africa to Europe at dusk, just before 6 pm. By 9 pm the Neanderthals had evolved in Europe and were manufacturing Levallois tools, while their counterparts in Africa were making similar technological advances.

At around 10.45 pm an early form of *Homo sapiens* arrived in the Middle East. By 11.20 pm Neanderthals had become the

Map showing major Neanderthal and modern human sites discussed in this chapter. The Danube River provided the first *Homo sapiens* to reach Europe with a route from the edge of the Black Sea to Germany and on to resource-rich southern France.

predominant type of human in western Asia and were steadily pushing eastwards towards Siberia, which they reached around 11.30 pm. By 11.40 pm, just twenty minutes before the present, all

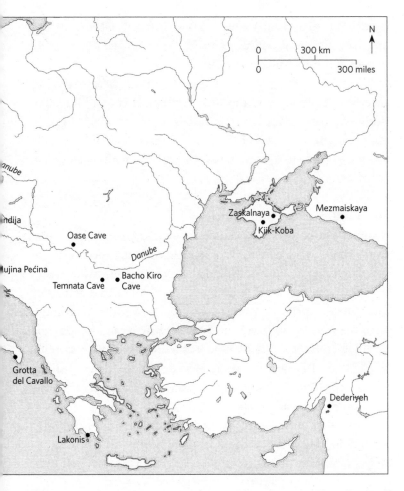

traces of the Neanderthals were gone. When the Last Glacial period ended at 11.54 pm, modern humans were on the verge of establishing long-term settlements and inventing agriculture.

In this chapter we will run through the principal suspects in the mystery of the Neanderthals' extinction. Were their bodies maladapted to a changing planet? Was their diet unsustainable or did it leave them vulnerable to competition? Did they live in too restricted a range of environments? Was their cultural life somehow deficient, indicating critical cognitive shortcomings? What was the impact of the arrival of *Homo sapiens*?

As we examine possible causes of their demise, we develop a new understanding of the Neanderthals. It is an unfortunate truth that extreme stress can be the best test of character. In learning about the character of the Neanderthals, who did not make it through these tough times, we also learn something about our ancestors who did.

Silver bullet extinction theories

Neanderthal extinction has become a media staple, an irresistible story where every new piece of evidence, no matter how small, gets magnified as a possible final answer to the age-old question. Yet when a silver bullet solution to the Neanderthal extinction conundrum is printed, it tends to be quickly overturned or simply forgotten. A review of recent news stories about Neanderthal extinction highlights these pitfalls. Readers should be especially sceptical when faced with any breaking news on this topic. All the headlines below are intriguing, but as a group they tend to cancel each other out – they cannot all be true.

Scientists and science journalists both have an incentive to exaggerate the importance of small discoveries. The press can attract a broader audience, while researchers can get good press coverage for their otherwise obscure project if they manage to tie it into the extinction question. Examples can relate to diet (Failure to Hunt Rabbits Part of Neanderthals' Demise?[1]), theories about social organization (Gendered Division of Labor Gave Modern Humans Advantage Over Neanderthals[2]), or seemingly small biological

---

1 *National Geographic News*, 12 March 2013

2 *Science Daily*, 5 December 2006

differences (Tolerance of smoke may have given us an edge over Neanderthals[3]; Neanderthals May Have Gone Extinct Due to Their Brain Shape[4]; Neanderthals had a propensity for earache, nudging them to their doom[5]).

The rabbit hunter theory, to the extent that it was ever taken seriously, lasted six years (Rabbit bones suggest Neanderthals were better hunters than we thought[6]). Theories of social or biological difference leave many unanswered questions. If gendered division of labour was putting Neanderthals at a disadvantage, surely they would have reorganized their work. And if a minor trait, such as swimmer's ear, threatened a whole species with extinction, then there would likely be a few members of that species who did not get swimmer's ear, and their DNA would quickly spread across the population.

Another common way of looking at the extinction question is to seek a catastrophic external cause, perhaps inspired by the asteroid that spelled doom for the dinosaurs. The complication here is that Neanderthals and modern humans both occupied Eurasia at the same time, and both populations would be confronted by the same challenge. Nevertheless, scientists look to the skies (Did UV rays spell doom for Neanderthals?[7]), to changing temperatures (Climate Change Likely Iced Neanderthals Out of Existence[8]) and to extreme weather triggered by a magnetic reversal (End of Neanderthals linked to flip of Earth's magnetic poles, study suggests[9]). As we discuss later in the chapter, the Phlegraean Fields in Italy erupted at a critical time. Could that be the solution? (Volcanoes Killed Off

3 *New Scientist*, 2 August 2016

4 History.com, 31 August 2018

5 *The Economist*, 12 September 2019

6 *New Scientist*, 6 March 2019

7 University of Florida News, 3 June 2019

8 *Smithsonian Magazine*, 28 August 2018

9 *The Guardian,* 18 February 2021

Neanderthals, Study Suggests[10]). Perhaps not (Supervolcano Cleared in Neanderthals' Demise[11]).

Computer models are a powerful way to look at how population changes might have occurred. In the hands of the science media, however, such models tend to be overblown as explanations in themselves, confusing the model with the cause (Declining fertility rates may explain Neanderthal extinction, suggests new model[12]). If fertility rates were going down, then that could obviously lead to extinction, but it does not explain why the rates were going down.

The most popular explanations of extinction are the most dramatic, in that they give a role to modern humans (Neanderthals may have died of diseases carried by humans from Africa[13]; Migrating humans may have killed off Neanderthals by accident[14]; Did interbreeding wipe out the Neanderthals?[15]; Did Dog-Human Alliance Drive Out the Neanderthals?[16]).

Few archaeologists would argue for any single cause, and yet these silver bullet theory headlines show no sign of fading. As we explore these causes, let us start with what we know about the Neanderthals just before their fall.

The classic Neanderthals

As we discussed in Chapter 1, the first bones formally identified as Neanderthals were the ones that emerged from the Feldhofer Cave in the Neander Valley in 1856. The confirmation that this represented a separate and unique population, or species (as opposed to a deformed individual), came from subsequent discoveries of a

---

10 *National Geographic*, 22 September 2010

11 *Live Science*, 24 October 2014

12 Phys.org, 29 May 2019

13 *The Guardian*, 11 April 2016

14 CNN, 15 April 2016

15 *Daily Mail*, 11 November 2018

16 *National Geographic*, 5 March 2015

pair of skeletons at Spy, Belgium, eight individuals at La Ferrassie, France, and a nearly complete skeleton at La Chapelle-aux-Saints, France. All of these Neanderthal individuals have been dated to between 70,000 years ago (the upper limit of the La Ferrassie dates) and 40,000 years ago (Spy and Saint-Césaire).

More recently, other nearly complete Neanderthal skeletons have been unearthed and dated to this same date range. The archaeologist Paul Pettitt compiled a list of Neanderthal skeletons that show some degree of intentional burial, and this list nicely illustrates just how rich the archaeological record is for the 'classic' Neanderthal period, in contrast to earlier times, when we were only able to discuss a handful of sites. From France, there is one individual from La Quina, two from Le Moustier, one from Roc de Marsal and one from Saint-Césaire (probably the most important one of the group for reasons we discuss below). In Syria, two children have been uncovered at Dederiyeh Cave. In Crimea, two Neanderthals

The skull of La Chapelle-aux-Saints 1, found in 1908 as part of the discovery of the first relatively complete skeleton of a Neanderthal (see also plate 13).

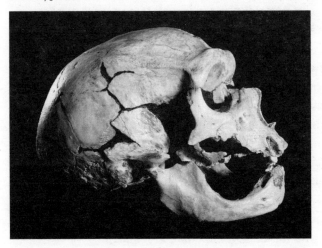

are from Kiik-Koba and a further three are from Zaskalnaya. In Russia, in the foothills of the Caucasus Mountains, a Neanderthal infant was excavated at Mezmaiskaya Cave.

We should add to this list two sites where the Neanderthal bones show signs of having been cannibalized: El Sidrón, Spain, was discovered in 1994 and contained remains of some thirteen Neanderthals, and Vindija Cave, Croatia, has several layers of deposits and was excavated extensively in both the early and late 20th century. These two sites have been key sources for extracting Neanderthal DNA.

There are other sites with Neanderthal remains from this period, but the ones listed above have been among the most important. In the late 1970s researchers started to establish formal criteria, mainly on the basis of the bones discovered in the late 19th and early 20th centuries, to define the Neanderthals. Albert Santa Luca and Jean-Jacques Hublin came up with four unique characteristics of Neanderthal skulls. These are all fairly technical features towards the back of the skull and seem to be related to the skull architecture needed to support extremely strong jaw muscles. On the other side of the skull, the absence of a protruding chin may also have related to chewing ability. Neanderthal faces were thrust forward and some researchers have likened them to the aerodynamics of racing cars. Such faces may have made it easier for Neanderthals to use their jaws as a vice, and heavy wear patterns on Neanderthal teeth support this explanation (see p. 78).

Jeffrey Schwartz and Ian Tattersall added a few more diagnostic criteria relating to the large Neanderthal nasal cavity. The Neanderthals had very large, broad noses, and anthropologists have offered competing and incompatible explanations for this, from an adaptation to cold climates (to warm the air during an intake of breath) to a cooling mechanism (to compensate for Neanderthals' high metabolism). Others think that the noses were just a chance variation with no adaptive explanation required.

El Sidrón Cave in northern Spain has yielded remains of thirteen Neanderthals showing signs of cannibalism, from which DNA has been extracted. They may have been members of an extended family.

Chris Stringer has noted other distinguishing features, including barrel-shaped chests, short limbs and two heavy brow ridges (shaped as two linked arches over the eyes). As with the nose, there seems to be no widely accepted adaptive explanation for these features. The brow ridges were perhaps their most distinctive feature in comparison to *Homo sapiens*. Brow ridges of one form or another have been prominent features of human anatomy since the days of *Homo erectus* (which had a single ridge like a unibrow), making modern *Homo sapiens*, with our smooth foreheads, the exception.

What do Neanderthal bodies tell us about them? From the basic structure of the bones, we know that they were extremely strong. From the extensive level of injuries found on the bones and the relatively young estimated ages at death for Neanderthals (few Neanderthal skeletons belonged to individuals who lived past the age of forty), we can surmise that life was hard for them. They probably fought animals at close range as part of the strategy of ambush hunting that developed back in *Homo heidelbergensis* times. And, indeed, it is likely that they fought each other, as all human groups do, except that they had an enhanced ability to inflict lasting damage.

Like us, they were mostly right-handed, and their arm bones show that their right arm muscles were stronger than their left arm muscles. But unlike us, their right arms were very much stronger

Vindija Cave, Croatia, is one of the sites to have produced Neanderthal bones from which DNA has been recovered.

than their left arms, a difference rarely seen today except among professional tennis players. It is likely that Neanderthals performed repetitive activities such as scraping hides for clothing, leading to a pronounced asymmetry in their body strength. It is possible that clothes-making through scraping was a significant burden on the Neanderthals' time. Bone needles are known only from modern human sites in later periods. However, the discovery in 2020 of 'fibre technology' (i.e. string) in Abri du Maras, France, indicates that Neanderthals did have the ability to attach stone tools to a shaft and make perishable artifacts such as rope, bags, nets and even basic clothing.

Neanderthals matured differently from modern humans. Chris Stringer and a number of collaborators have looked at Neanderthal teeth, which leave growth marks not unlike tree rings. They argue that Neanderthals matured much faster than *Homo sapiens*. For example, a Neanderthal child from Le Moustier showed a level of maturation of a modern sixteen-year-old but has been estimated to have been only twelve years old. On the other hand, Antonio Rosas and his colleagues looked at an eight-year-old Neanderthal boy from El Sidrón and argue that his growth rate was similar to ours today, although they say his brain was developing at a slightly slower rate. This child, however, had already worn his teeth down significantly, meaning that he was carrying out chores in hide preparation.

Putting this together, Neanderthals grew up fast – or at least took on adult responsibilities at an early age – and died young. Most knew combat, most likely with large animals. Despite these hardships, they were a caring people. Some individuals – dating back to the Sima de los Huesos and continuing into the classic Neanderthal period at Shanidar – had suffered debilitating injuries (which we can see in their partially healed bones) and would have needed support in order to survive.

Many Neanderthal features, including their heavy build and stocky bodies, have long been interpreted as adaptations to a cold climate. As part of human variation today, there is a correlation between the latitude where a population lives and the stockiness of their bodies. The explanation for this is that the human body is better adapted to cold when its surface area is limited by shortened limbs, as this helps to conserve heat. This is known as Allen's rule after the 19th-century Harvard zoologist Joel Asaph Allen. Another such rule is Bergmann's rule, named after the 19th-century German biologist Christian Bergmann, which correlates cold temperatures with large overall body sizes.

Following these rules, Neanderthal bodies reflected their long exposure to cold compared with the lankier *Homo sapiens*. But the timing of the Neanderthal extinction is something of a paradox. Their bodies had adapted to cold during a previous severe glaciation. Around 37,000 years ago the European climate started another slow decline, culminating in the Last Glacial Maximum some 26,000 years ago. By the time of this peak glaciation the Neanderthals were already extinct. Why did they die out during a non-severe cold period during which the more tropical-adapted *Homo sapiens* were thriving? Their physical adaptations were clearly not giving them an edge.

Researchers have estimated that Neanderthal thermoregulation would have given them only a small advantage over their *Homo erectus* forebears or *Homo sapiens* rivals, amounting to just 1°C increased tolerance to cold. During European winters and in glacial periods this would have been of little help, and the Neanderthals would still have been dependent on good clothing, fire and possibly a high-fat diet to stay warm. Yet their body mass and proportions came at a price. In comparison to the more gracile *Homo sapiens*, the Neanderthals would have needed more calories to support their muscular bodies. How did they find these large meals? To answer this, we turn to Neanderthal diet.

There are two ways to reconstruct Neanderthal diet. One is to look at the animal bones discarded at Neanderthal occupation sites. This method can tell us much about what species the Neanderthals ate, whether they selected game in their prime or settled for eating the old and infirm and in what seasons they tended to butcher animals. But this method is far from comprehensive, for it excludes plants, the remains of which do not survive all that well, and large game which Neanderthals probably consumed at kill sites. There is the additional problem that animal carcasses may have been brought into caves by other carnivores.

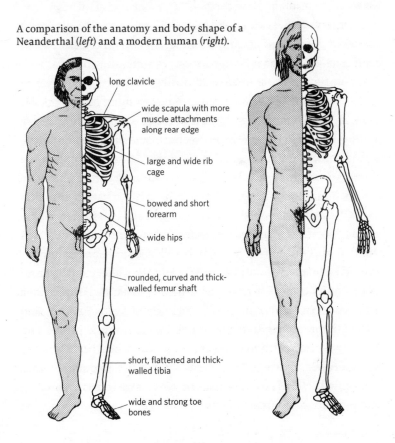

A comparison of the anatomy and body shape of a Neanderthal (*left*) and a modern human (*right*).

long clavicle

wide scapula with more muscle attachments along rear edge

large and wide rib cage

bowed and short forearm

wide hips

rounded, curved and thick-walled femur shaft

short, flattened and thick-walled tibia

wide and strong toe bones

To compensate for these shortcomings archaeologists are able to learn more from the ratio of certain isotopes in ancient bones and teeth. By extracting bone collagen (which is produced from dietary protein) from Neanderthal remains and looking at the ratios of stable carbon and nitrogen isotopes, researchers can get a picture of what kinds of protein they were consuming. Meat eaters tend to have different isotope ratios from herbivores, and archaeologists can compare the isotope ratios in Neanderthal collagen to the ratios found in other animals from the same area. This method, like carbon dating, involves the destruction of precious remains, however small, of the Neanderthals, so archaeologists are reluctant to do it too often. Luckily, stable isotope ratios can be discerned as part of the process of carbon dating, so both procedures can be conducted by the same lab on the same material.

Combining these lines of evidence, a picture has emerged of the Neanderthals as top predators, consuming huge amounts of meat, particularly from large mammals such as woolly mammoths. They also consumed horse, red deer, reindeer and bison. A wealth of habitation sites in France – most notably Combe Grenal (Dordogne), which was excavated by François Bordes in the 1950s and 1960s and which has 13 metres (40 ft) of occupation levels dating from more than 125,000 years ago to less than 50,000 years ago – has shown that Neanderthals were primary butchers (i.e. they did not scavenge animals killed by other predators or animals who died of natural causes) and they selected prey from throughout herds, rather than selecting the weak. There has been a long debate over this issue, led on one side by Lewis Binford, who argued that the Neanderthals were simple scavengers. There is no doubt that Neanderthals did scavenge for food, but the view of the Neanderthals as expert hunters has lately come to predominate.

A new kind of evidence has emerged more recently, showing that the Neanderthals ate more than just meat. The anthropologist

Karen Hardy studied chemicals in the plaque of Neanderthal teeth from El Sidrón on the north coast of Spain, and found evidence that they were eating roasted vegetables and consuming chamomile, possibly for medicinal purposes. A similar study of teeth from Shanidar, Iraq, and Spy, Belgium, showed evidence for the consumption of plants, including wild grains. A study of bacteria on ancient teeth found that both Neanderthals and early modern humans had an oral microbiome that specialized in freeing sugars from starchy foods, indicating that carbo-loading was an underappreciated part of Neanderthal subsistence. They were likely eating a lot of nuts, root vegetables, probably cooked, and in some cases cooked grains. Coprolites (fossilised poop) from a cave in Spain also show that while the Neanderthals ate a lot of meat, they also clearly ate plants. This diet is much more varied than that implied by the isotope data alone, although it does not contradict the notion that Neanderthals were heavily meat-dependent.

For archaeologist Clive Gamble, 'The isotope revolution has cut out all the endless speculation about scavenging and hunting and returned the verdict they were top predators, albeit ones who did not eat much fish, something that the earliest moderns in Europe did.' The consumption of fish has emerged as perhaps the greatest dietary contrast that Neanderthals had with the *Homo sapiens* that replaced them in Europe.

There is one part of Europe where the climate, environment and food resources were vastly different from areas such as France where we have built up this picture of Neanderthals as serious meat eaters. We refer to the area called a 'Mediterranean Serengeti' by Clive Finlayson, the director of the Gibraltar Museum, who has spent more than a decade studying Gorham's Cave (see plate 2) and other Neanderthal sites in Gibraltar. Finlayson argues that the food available to Neanderthals in Spain in this period was more diverse than the food they found in the heart of Europe.

The Neanderthals in Gibraltar, from the Eemian interglacial until their extinction, ate rabbit along with grazers such as ibex and red deer. In addition they consumed mussels, limpets, cockles, tortoises, monk seals and, incredibly, dolphins. Like their modern human counterparts in Africa, their consumption of molluscs goes back to the earliest levels, perhaps before the Eemian. Sites from around the Mediterranean and the Near East show similarities with Spain in that the Neanderthals were eating small game such as red deer, along with tortoises, seafood and plants.

Recall from Chapter 5 that early modern humans in Africa seem to have made their great behavioural leap forward in a coastal environment, and there is a wealth of key sites in Morocco and South Africa. *Homo sapiens* at the time seemed to recognize the importance of the sea, as some of the world's first symbolic behaviour involved the creation of jewelry out of mollusc shells. It is certainly possible that *Homo sapiens* were the beneficiaries of the particular geography of Africa, with its long coastlines adjacent to grasslands that supported herds of game.

In contrast, the geography of Europe is such that the main coastal areas were in less productive regions in the south. The Mediterranean coast is also less productive than the South African coast where some of the earliest modern behaviours appeared. In Spain, at least, the Neanderthals were learning to exploit the coast and take advantage of its rich resources. In 2020 news came of a coastal cave in Portugal called Figueira Brava, where Neanderthals were eating seafood as early as 106,000 years ago. The problem is that as part of the Iberian Peninsula, the coast of Spain is limited, and the Pyrenees, especially in colder times, acted as a barrier to the rest of Europe. The Neanderthals were more or less locked into an area only a small fraction of the size of coast that *Homo sapiens* exploited in Africa. There is evidence that Neanderthals consumed shellfish in Italy too and seem to have been diving to collect clam shells (found at the site of Grotta dei Moscerini) to use as tools, but as in Spain and Portugal they were not able to expand their numbers or range in a significant way.

Some researchers have presented compelling evidence that the Neanderthals were doing more than harvesting seafood in the Iberian Peninsula. Finlayson and others have proposed that in Gibraltar and other areas Neanderthals were plucking feathers and removing talons from inedible birds such as eagles, possibly for ornamental purposes. In 2015, archaeologists Davorka Radovčić and Ankica Oros Sršen argued that Neanderthals used eagle claws decoratively at Krapina, Croatia. Similarly, in 2017 the Italian archaeologist Francesco d'Errico published the discovery of a notched raven bone from the Neanderthal site of Zaskalnaya in Crimea, providing support for the notion that Neanderthals were interested in birds for symbolic reasons. Further evidence has emerged for even more sophisticated symbolic behaviour. At Gorham's Cave in 2014, Finlayson and his team discovered a cross-hatch design etched on the cave wall. And in 2019, German excavators found abstract-looking carvings on a knucklebone of a large animal, which they dated to be older than the radiocarbon limit of 50,000 years ago, putting it into Neanderthal times.

More controversially, the Portuguese archaeologist João Zilhão has argued that Neanderthals were experimenting with cave art and creating paints of different colours. His claim is based on U-series dating, a technique that measures the amount of decay of radioactive uranium and thorium on deposits overlaying prehistoric cave paintings. His team pointed to possible art in three Spanish caves: geometric shapes at La Pasiega, a hand stencil (made by blowing paint on a hand held against the wall in order to create an outline) at Maltravieso and lines at Cueva de Ardales, all with minimum dates around 65,000 years ago. A subsequent study in 2021 confirmed these dates. Another cave, El Castillo Cave in

Cantabria, Spain, has even more sophisticated art: a series of red dots and hand stencils. This site is dated to over 40,000 years ago (see plate 10) and could be the work of Neanderthals or could represent the arrival of a new population.

With the same method Zilhão argues that perforated shells and pigments from an additional site, Cueva de los Aviones, are more than 115,000 years old. These dates remain under dispute. Corroborating evidence for such early Neanderthal symbolic behaviour came from France in 2016, when investigators found a circular ring of stalagmites in Bruniquel Cave, dating to 175,000 years ago. Neanderthals used red ochre from as far back as 250,000 years ago, initially for its use as a dye and for its adhesive qualities. For Chris Stringer, one of the biggest surprises in his career so far has been the 'recent evidence of the [Neanderthals'] use of pendants, pigments and adhesives to make composite tools'. While it is not entirely clear that the Neanderthals made the leap to symbolic behaviour, it seems that if they did, this phenomenon, like fishing, may not have been widespread.

What the DNA tells us

Increasingly, DNA evidence can provide a fuller picture of what the Neanderthals were like, while also shedding light on the circumstances around their extinction. Perhaps the most important finding is the degree of homogeneity in the global Neanderthal population. Our own species is unusual, compared with other primates, in that we are also very homogenous. Yet Neanderthals and Denisovans had far less genetic variation than present-day modern humans have, even outside Africa. 'Neanderthals have off-the-charts low diversity,' says Harvard geneticist David Reich.

The first clue to Neanderthals' low variability came in the year 2000, by which time there were three published mtDNA sequences (from the Neander Valley, Vindija and Mezmaiskaya). According to

Pääbo's calculations, the differences were quite small, meaning that all three individuals had a relatively recent common female ancestor.

More recent studies and advances in DNA sequencing have shown that Neanderthal genetic diversity was contracting around 50,000 years ago. There was also inbreeding. A high-quality genome from the toe bone of a Neanderthal woman from Denisova Cave in Siberia (the first Neanderthal genome ever sequenced to high quality) revealed that in several parts of her genome the two strands of her DNA were identical. Her parents must have been closely related, maybe even half-siblings. Inbreeding was also common among her recent ancestors, usually a sign of small groups living in relative isolation. The second high-coverage genome was from Vindija Cave, Croatia, where the first draft Neanderthal genome had been produced in 2010. The third high-coverage Neanderthal genome was from Chagyrskaya Cave in 2020, which is also in the Altai Mountains, and it confirmed the result from nearby Denisova Cave, that the Neanderthals in this region lived in small groups with heavy inbreeding.

In 2021, Pääbo and his colleagues Benjamin Peter and Laurits Skov were able to compare the genes of a dozen individual Neanderthals, most of them from Chagyrskaya Cave. What they found was that the men were closely related, indicating that these Neanderthals were patrilocal. This means that men tended to stay with the family while women would leave their home group. The study also showed that the male genetic diversity was at dangerously low levels. There were small numbers of males contributing to reproduction. It could be that this problem was particularly severe for the Siberian Neanderthals, living so far from the more populated areas in the west. Or maybe Neanderthals were simply heading toward extinction.

What does DNA tell us about Neanderthals' looks? Neanderthals share modern human genes for green eyes and red hair. As many had predicted, given their evolution in Europe, the Neanderthals seem to have evolved genes for pale skin. Amazingly, genetics also tells us that Neanderthals had a propensity to smile in order to show friendliness and perhaps win sexual partners.

For their abilities, the genes confirm that Neanderthals had some form of speech. They also had a gene called FOXP2, which is the same as it appears in modern humans. FOXP2, initially heralded as 'the language gene', is related to the fine motor skills, coordination and executive function required for producing complex sounds for speech and, presumably, also for making complex stone tools. This implies that both species inherited FOXP2 from a common ancestor.

The science of ancient DNA is still very young, and there will be more to learn about Neanderthal appearance and behaviour. According to Stringer, 'genomic work will lead the way, once we can achieve high quality reconstructed Neanderthal genomes to compare with those of chimps, humans and Denisovans. We will start to tease out features that typify each lineage, and at least make a start in looking at possible differences in brain function'. Much progress has been made since the first edition of this book, but we have only just scratched the surface.

Enter the Aurignacians

In 2012 bone flutes and portable art such as human figurines, some with animal heads, from the site of Geissenklösterle, Germany, were dated to more than 42,000 years ago. Similar art from other sites in the Swabian Jura, such as Hohle Fels, are dated to almost that old. With Geissenklösterle and Hohle Fels, we have the first examples of figurative art in Europe, and this represents a significant new development. The only other example in the world where representational art, including human-animal hybrid images, has been found from this time is in cave paintings on the Indonesian island of Sulawesi.

For many years carbon dating could not produce reliable dates approaching 40,000 years ago. With the introduction of the technique called ultrafiltration, however, there has been a great improvement in reliability. Now it is possible to carbon date smaller samples. This limits the effect of contaminants in the form of more recent carbon isotopes, which can make organic material appear to be much younger than it is. The smaller samples also mean that museum curators are more likely to allow direct dating on valuable finds such as worked animal bones, shell beads and human bones, because the destruction is more limited.

A similar revolution has taken place in U-series dating. With less destruction caused in the collection of small samples, it is now possible to use the U-series technique – which can go beyond the limits of carbon dating – to date cave paintings. Thanks to ultrafiltration and the new U-series dating, much of the first portable and cave art in Europe, and elsewhere in the world, has been pushed back in time.

The figurative portable art and the bone flutes are part of a suite of changes seen in Europe from just before 40,000 years ago and constitute an industry known as the Aurignacian, from the name of a cave in the Midi-Pyrénées region of France. Paul Mellars has named six elements which together comprise the Aurignacian and distinguish it from the Mousterian, the industry that had been favoured by the Neanderthals since 250,000 years ago. These

A flute made from a bird bone found at Geissenklösterle, Germany, and dated to more than 42,000 years ago, signalling the arrival in Europe of modern humans.

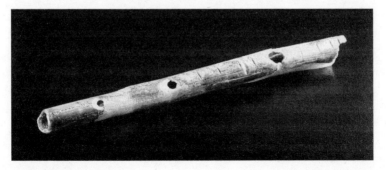

are: (1) improved blade production using soft-hammer percussion (see our discussion of stone tools in Chapter 4); (2) more sophisticated stone tools with an emphasis on blades; (3) the use of tools made of bone, antler and ivory as well as stone; (4) ornaments (shell beads); (5) art; and (6) expanded trade networks.

The appearance of the Aurignacian represents such a radical departure from the preceding 200,000-plus years of Mousterian toolmaking that archaeologists mark this moment as the end of the Middle Palaeolithic and the beginning of the Upper Palaeolithic period. This change is also conventionally seen as the end of the Neanderthals and the beginning of the time of modern humans. The actual transition is fuzzier than that, and the question of whether Neanderthals were capable of creating Upper Palaeolithic tools has been fundamental to the debate over their cognitive capabilities.

With the exception of a few dissenters, such as Zilhão and d'Errico, who believe that Neanderthals made some of the advances associated with the Aurignacian on the eve of the arrival of *Homo sapiens*, most researchers credit the entire suite of Aurignacian behaviours to newly arrived modern humans. Who were these *Homo sapiens* and how did they get to Europe?

The sudden and widespread appearance of modern humans in Europe over 40,000 years ago requires some explanation. As archaeologists try to trace the spread of *Homo sapiens* around the world, perhaps the most important line of evidence to emerge in recent years comes from DNA. Though there seems to be a great deal of variation among human populations outside of Africa, the DNA tells a different story, which reveals that most current human variation can be found within Africa. In contrast, the rest of the global population is descended from a founder group that could have been as small as just a few hundred individuals.

Archaeologists and geneticists are still working out the details of this extraordinary story of global colonization. Consensus is building that a main African exodus event occurred around 80,000 to 65,000 years ago, either across the Bab-el-Mandeb ('Gate of Grief'), roughly from modern Djibouti in the Horn of Africa to Yemen in the Arabian Peninsula, or alternatively around the rim of the Red Sea. From that point the spread around the world occurred with incredible speed. Research in 2017 dates the human arrival in Australia at 65,000 years ago, near the lower estimate for the start of the dispersal.

What accounts for the rapid spread to Australia in contrast to the time it took modern humans to reach Europe? Even today the shoreline has retained a central place in the social and economic lives of *Homo sapiens*. It is a place of recreation and seafood harvesting. As an essentially one-dimensional geographic zone (a thin line between scrub and shore), it lends itself to safe forward exploration (one can always double back) and social gatherings. For beachcombing pioneers coming out of Africa, one can imagine it as a very long beach holiday lasting thousands of years, where each generation could move a little further along the coast into empty land.

A population expanding around the southern coast of the Arabian Peninsula could have reached India and beyond quite quickly. The coastal route to Australia, despite its long distance in absolute terms, is easier ground to cover than the route inland. Of course, not everyone stayed on the beach. The pioneers who made their way to Europe could have come by any of a number of routes – directly north from Yemen, through the Fertile Crescent in Iraq or via a more circuitous path. Most of these routes would have brought them to Turkey, Bulgaria and Romania, where they would have discovered the Danube as a sort of prehistoric European superhighway to central Europe (another essentially one-dimensional geographic zone), exactly where some of the earliest European portable art has been found.

At the gateway to Europe, the earliest modern human sites are Oase Cave, Romania, which was discovered in 2002, and nearby

Temnata Cave and Bacho Kiro Cave in Bulgaria. More controversially, carbon dating with ultrafiltration on modern human teeth from old excavations has given dates of more than 41,000 years ago for Kents Cavern, Torquay, in England (Britain was then accessible by a land bridge), and more than 43,000 years ago from Grotta del Cavallo, Apulia, Italy. In addition, a possible modern human tooth fragment from Grotte Mandrin in France has been dated to around 54,000 and could represent an earlier and less successful foray.

A fairly complete *Homo sapiens* skull from Oase has a curious blend of modern and archaic features. In 2015, ancient DNA from the jawbone of a second individual from Oase has been shown to have more than 6 per cent of its DNA from a recent Neanderthal ancestor, and this is a much larger percentage than one finds in people living today. Jawbones from Bacho Kiro have been the subject of debate as to their categorization. In 2021, DNA studies of three individuals from Bacho Kiro found that they were *Homo sapiens* who had Neanderthal ancestry within the previous five-to-seven generations. Further along the Danube corridor, just south of Prague, Czech Republic, is the site of Zlatý kůň, where a skull that could represent the oldest modern human found in Europe has been shown to have Neanderthal ancestry too, but much deeper in the past (over 60 generations). These bones make it clear that there were multiple waves of modern human arrivals into Europe and that by the time they arrived, they had already intermixed with Neanderthals in the Middle East.

The cultural explosion that goes by the name Aurignacian is associated with the intrusion of modern humans, albeit with some Neanderthal intermixing, into Europe. At the site of Fumane, in northern Italy, a Protoaurignacian assemblage includes human teeth with modern human mtDNA. This brings us to the next suspect in our list of possible causes of the Neanderthals' demise. Did the Aurignacians wipe them out? The key to answering this question is to look at population patterns through time. Let us start with the coast of south-eastern Europe, which is sometimes proposed as a possible route into the continent by Aurignacian pioneers. Dimitra's post-doctoral research addressed exactly this question. The evidence for overlap between Mousterian and Aurignacian populations can be challenging. At many sites there is extensive mixing from different time periods, and collections from a single level may have accumulated over tens of thousands of years. Researchers tended to use non-committal terms for mixed Mousterian/Aurignacian archaeological sites, calling them Terminal Middle Palaeolithic, Transitional Middle/Upper Palaeolithic, Initial Upper Palaeolithic or Early Upper Palaeolithic. All these terms amounted to the same thing, which is that the collections of stone tools were not easily classified as Mousterian (and probably Neanderthal) or Aurignacian (and probably modern human).

Dimitra argued that the youngest radiometric dates associated with Mousterian tools, from such sites as Mujina Pećina, Croatia,

Excavation at Mujina Pećina Cave, Croatia.

and Lakonis Cave, Greece, place them at more than 40,000 years ago. Meanwhile, the earliest Aurignacian sites from the region were dated to just 37,500 years ago. To Dimitra this indicated that the mixed tool collections were not evidence for overlap. She argued instead that the Aurignacian culture may have arrived late on the Balkan coast. By then the local Neanderthal population had been long gone, and the region had been empty of human habitation.

Is this pattern of hiatus between Neanderthal departure and modern human arrival observable elsewhere? In the Caucasus Mountains, the Neanderthal infant at Mezmaiskaya Cave, Russia, was once thought to be a late Neanderthal survivor at around 35,000 years old. In 2011, however, the infant was re-dated to about 40,000 years ago, before modern humans arrived in the area. The Caucasus region is now considered to have been devoid of Neanderthals by the time modern humans arrived there. The ultrafiltration revolution in carbon dating is pushing many late Neanderthals to earlier than 40,000 years ago, throwing new doubt on the question of overlap between the two species in Europe.

Another region to be a subject of debate is Italy. A localized stone tool tradition in southern Italy known as Uluzzian has long been considered 'transitional' from Middle to Upper Palaeolithic, basically because it is not Mousterian and not Aurignacian, and the hominin bones associated with it were not definitely categorized as Neanderthal or modern human. The Uluzzian is known for having crescent-shaped knives with extensive retouching. No one could say for sure whether the toolmakers were Neanderthals who had 'progressed' from the Mousterian, or modern humans who had not yet achieved the full Aurignacian suite of technologies, so 'transitional' seemed to cover it.

Until recently the conventional wisdom was that the Uluzzian represented a limited adaptation of Neanderthals to Upper Palaeolithic tool forms with which they had come in contact. But in 2011 two teeth at the Uluzzian site of Grotta del Cavallo – the first Uluzzian site identified in the 1960s – were identified as modern human and, as we mention above, dated to more than 43,000 years ago. While some dispute these dates, the Uluzzian is now considered by many to be a modern human tool tradition, and this has removed the evidence that Neanderthals survived in Italy well after the appearance of modern humans elsewhere in Europe. Italy now seems like a mosaic, with the Uluzzian and Aurignacian (probably modern humans) and Mousterian (probably Neanderthals) in different regions at different times.

While there are continuing disputes over new technologies and particular dates, it is widely accepted that Neanderthals and modern humans shared Europe from the modern human arrival at least 45,000 years ago until the Neanderthals' extinction. But this does not mean that they were together for that whole time in all places. New dates and new dating techniques are appearing with great frequency, and the continental picture is rapidly changing. With that caveat in mind, what we can say is that the Neanderthals had already retreated from many areas well in advance of the arrival of modern humans. But this was not necessarily true everywhere.

There is, in fact, some overlap between the Mousterian and Aurignacian (and presumably Neanderthal and modern human) occupation in northern Spain and in France. As the dating is refined, the period of overlap has gone down in researchers' estimates from as much as 10,000 years to perhaps just a few thousand years.

There is little consensus over the nature of interaction between Neanderthals and *Homo sapiens* in these areas. Some claim that over thousands of years the two populations were so small and the population densities so low, perhaps as little as one person per 100 square kilometres (40 sq. miles), that they may have missed one another. For others, their direct interactions were likely to have been

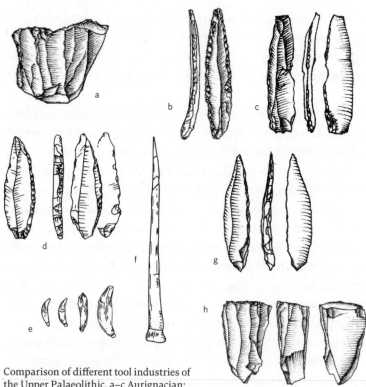

Comparison of different tool industries of
the Upper Palaeolithic. a–c Aurignacian:
prismatic core (a: length approx. 46 mm),
retouched blade (b: length approx. 95 mm) and retouched bladelet (c: length
approx. 31 mm), from Cosava, Romania; d–f Châtelperronian: points (d: length
approx. 64 mm), perforated animal teeth ('pendants') (e: greatest length approx.
46 mm) and bone artifact (f: length approx. 112 mm), from the Grotte du Renne,
Arcy-sur-Cure, France; g, h Gravettian: backed bladelet (g: length approx. 30 mm)
and prismatic core (h: length approx. 42 mm) from Bistricioara-La Mal, Romania.

frequent. Until recently there was a widespread belief that at least
one group of Neanderthals in southern France became 'accultur-
ated' to the Upper Palaeolithic technology of their modern human
neighbours. This group is associated with an industry known as the
Châtelperronian, which has long been at the centre of debate over
the Neanderthals' cognitive abilities.

The Châtelperronian culture was first identified at a site in Auvergne in central France. It gained notoriety thanks to André Leroi-Gourhan, one of the grand figures in the history of French archaeology. Leroi-Gourhan was instrumental in preserving some of the treasures of the Louvre during the Second World War. He also pioneered large-scale horizontal excavations of prehistoric sites, in order to reveal activity areas and site organization rather than focusing on stratigraphy. From 1949 to 1963 he excavated at the Grotte du Renne, a cave at Arcy-sur-Cure in Burgundy. The stone tools he found, like the Uluzzian, seemed to be intermediate between Mousterian and Aurignacian types. Blades predominated, with extensive retouching on a long, curved surface that was intentionally dulled, presumably so it could be held in the hand for cutting. These were probably made with soft-hammer percussion, making the Châtelperronian an Upper Palaeolithic industry.

Leroi-Gourhan also found ivory ornaments and pierced teeth at the Grotte du Renne. These artifacts were symbolic and were unlike anything that had been associated with the Neanderthals. Yet Leroi-Gourhan uncovered Neanderthal teeth within the Châtelperronian layers. These indicated that something extraordinary was happening here. Specialists such as François Bordes and Paul Mellars argued that Châtelperronian stone tools evolved out of a local Mousterian tool variant, the Mousterian of Acheulian Tradition (which was the one Mousterian variant that Bordes considered to be chronologically restricted and late in the sequence). The ornamental ivory and teeth seemed crudely made, as if by humans who had seen jewelry but who were unskilled at its manufacture and may not have fully understood its purpose. In 1979 this line of argument received a huge boost in the form of a Neanderthal skeleton unearthed among Châtelperronian artifacts at Saint-Césaire, near the Atlantic coast. At the same time, the Saint-Césaire skeleton (see plate 12) helped put to rest another theory, which was that Neanderthals evolved into modern humans in Europe. The presence of a 'classic' Neanderthal skeleton from just 40,000 years ago among Upper Palaeolithic tools was strong evidence that modern humans were intrusive and evolutionarily separate from Neanderthals.

Today the tide is turning once again. Re-dating of the Châtelperronian layers at the Grotte du Renne using ultrafiltration has shown evidence of mixing, with material as old as 49,000 years ago (before the start of the Châtelperronian) and as young as 21,000 years ago (after the extinction of the Neanderthals). Researchers are also questioning the association of the Saint-Césaire Neanderthal and Châtelperronian tools. It now seems that the skeleton is in a mixed layer not unlike the 'Transitional Middle/Upper Palaeolithic' layers in Greece, casting doubt on its very late date. A further blow to the acculturation argument is that stone tool experts such as Jean-Guillaume Bordes (no relation to François) are now questioning whether the Châtelperronian evolved from the Mousterian.

Drawing of the skull of a 'classic' Neanderthal found at Saint-Césaire, dated to around 40,000 years ago (*left*) and the skull and mandible (not from the same individual) of an Upper Palaeolithic modern human unearthed in the Czech Republic (*right*).

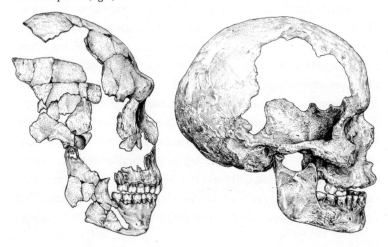

If the Châtelperronian turns out to have been a modern human industry, is this an argument against Neanderthal cognitive abilities? Yes and no. While the Châtelperronian was once seen as a Neanderthal 'acculturation' to the Upper Palaeolithic, it was always seen as inferior to the neighbouring Aurignacian tradition. It had defined the limits of what Neanderthals could achieve. If we now accept that there is no solid evidence for Neanderthal use of Upper Palaeolithic technology, we can no longer say that the Neanderthals could only effect a poor imitation of the Aurignacian. This leaves us wondering, if the Châtelperronian is a modern human industry, would it have always been considered 'inferior' to the Aurignacian?

Can we blame the Neanderthals' extinction on the modern human pioneers entering Europe? At this point, with lingering uncertainties about radiometric dates, the circumstantial evidence is strong but the case is not proven. In many areas, such as the Caucasus Mountains, southeastern Europe and parts of Italy, *Homo sapiens* entered territory that had been vacated by Neanderthals. In other territories, such as France, southern Germany and northern Spain, we have a picture of two separate populations living side-by-side with possible localized acculturation. With DNA evidence of admixture proving that Neanderthals and modern humans interbred for thousands of years before extinction, we are left wondering what had changed by 40,000 years ago? Perhaps this marked the end of a long process of decline. Perhaps the Neanderthals faced a myriad of challenges, and the appearance of modern humans represented the final blow they could not withstand.

As this process unfolded, the Phlegraean Fields volcano in Italy erupted 39,000 years ago (see pp. 180–83). Evidence for the continuing survival of the Neanderthals after that time is controversial. But they were not the only humans to disappear around then. The Châtelperronian and Uluzzian traditions, both possibly modern human, ended along with the Mousterian. The Aurignacian would not last much longer. A new modern human culture, the Gravettian, would soon sweep across Europe.

THE IMPACT OF VOLCANOES

Some time between about 71,000 and 75,000 years ago a supervolcano on Mt Toba on the island of Sumatra, Indonesia, erupted in perhaps the most massive explosion the world has experienced during the whole course of human evolution since the genus *Homo* first appeared in Africa. The Toba eruption had effects that were far-reaching, both geographically and through time. Closer to the Neanderthals' home, and to the crucial millennia around their extinction, an eruption near Naples at the Phlegraean Fields 39,000 years ago was not quite as big but could well have triggered a 'volcanic winter' that lasted many years.

Could these two volcanic events have been the true culprits behind the extinction of the Neanderthals, with *Homo sapiens* no more than opportunists who swept in afterwards? The Toba super-eruption ejected as much as 3,000 cubic kilometres (700 cu. miles) of magma. Debris has been identified as far away as India. It is likely that most mammals and birds within 350 kilometres (200 miles) of the eruption would have been instantly destroyed. Global temperatures dipped for several years and ecological disruption lasted a few hundred years. What was the consequence of this for humans?

Indonesia is far from the centres of human life in this period. Neanderthals were restricted to Europe and western Asia, and modern humans were just starting – or just about to start – their major exodus from Africa. The island of Flores, where the dwarf-like *Homo floresiensis* has been

Satellite view of Lake Toba, Indonesia, the remains of a supervolcano that erupted at some point between 71,000 and 75,000 years ago, possibly the largest volcanic eruption since the dawn of humanity. Could this or a smaller eruption at the Phlegrean Fields, near Naples, Italy, 39,000 years ago, have played a part in the extinction of the Neanderthals?

found, was outside the 350-kilometre kill zone, but would nevertheless have been under extreme stress.

The archaeologist Michael Petraglia, working at Jwalapuram in India, has argued for continuity in stone tool manufacture before and after the Toba eruption. The intriguing question about the tools – and Petraglia has found only a few hundred artifacts both above and below the Toba ash line at Jwalapuram – is who made them. He believes they were made by modern humans, while others, citing genetic evidence, believe that modern humans did not leave Africa until later.

The mystery of the Indian Middle Palaeolithic highlights the fact that the various hominin species in Asia all used similar stone technology in this period, making it hard to separate Neanderthals, Denisovans and modern humans in the absence of human remains. In any case, it seems whoever was making stone tools in India around the time of the Toba eruption survived the ordeal.

When most of us think of volcanoes near Naples, Italy, we think of Mt Vesuvius, which buried the Roman city of Pompeii. Some 20 kilometres (12 miles) further west along the coast is the Bay of Pozzuoli, which comprises part of another caldera. It is this volcano, known as the Phlegraean Fields, that erupted 39,000 years ago, when modern humans were already established in Europe and the Neanderthals' fate was uncertain. This eruption threw out 250 cubic kilometres (60 cu. miles) of magma, a fraction of the size of the Toba super-eruption, yet ash from the volcano – detectable only through microscopic particles of glass showing the chemical signature of the event – reached Asia.

In the years around the Phlegraean Fields' eruption a large number of icebergs entered the north Atlantic, cooling global temperatures in what is known as a Heinrich event. This was a difficult time, to be sure.

The evidence of these two volcanic events serves as a worrying reminder of the fragility of our environment. The second of these eruptions took place when the Neanderthals were vulnerable, as modern humans had already encroached on their territories. It is possible that it could have been the final blow, quickening a process that was long under-way. We know that modern humans survived the blast, thereby showing a degree of resilience – or perhaps merely a demographic advantage in their stronger presence in Asia

– that the Neanderthals may have lacked. It may be that the Neanderthals and the earliest modern humans in Europe were in the wrong place at the wrong time.

Exit, Stage 3

What was the climate of Europe like during these eventful years? Could a glacial period explain the Neanderthals' disappearance?

These questions were at the heart of a major interdisciplinary research initiative called the Stage 3 Project, which was the brainchild of Sir Nicholas Shackleton and Tjeerd van Andel at Cambridge. Stage 3 refers to the geological period from 60,000 to 24,000 years ago, when the ratio of oxygen isotopes, measured in deep-sea cores, indicates an interglacial or warm period. Stage 2 included the Last Glacial Maximum, popularly known as the peak of the Ice Age, and we currently live in Stage 1, an interglacial known as the Holocene.

Stage 3 was not nearly as warm and stable as Stage 1 (which started between 11,000 and 12,000 years ago), or as the Eemian (130,000 to 120,000 years ago), which we discussed in the previous chapter and which took place during Stage 5. Instead, Stage 3 was incredibly variable, which was itself a major source of stress. Temperatures fluctuated between warm and cold, with some cycles as short as 1,000 years. During the warm periods the climate would have been quite mild. For example, 45,000 years ago the mean temperature in July in north-eastern France is estimated to have been between 16°C and 22°C (61–72°F). Today that number is around 20°C (68°F).

Climate records show that northern Europe remained habitable during most of Stage 3. The lowlands of Scandinavia were not covered by glaciers until as late as 26,000 years ago. The persistence of warm conditions points away from cold climate alone as the cause of the Neanderthals' extinction, as their numbers had gone down well before the climate reached its most challenging state. On the other hand, the general warming of Stage 3 provided an opportunity for modern humans to expand into Europe.

One way to investigate the effect of the fluctuating climate on the Neanderthals is to look at its impact on other animals. Many large mammals, such as reindeer, horse, steppe rhinoceros and mammoth, as well as smaller species, such as the wolverine, wolf and fox, all survived Stage 3 and were present during the Last Glacial Maximum. Mammals that did not survive in Europe include the leopard, certain species of weasel and marten, and large species such as Merck's rhinoceros and the straight-tusked elephant. There are intriguing parallels in the patterns of retreat and extinction between the Merck's rhinoceros, straight-tusked elephant and the Neanderthals. It could be that the Neanderthal population was shrinking along with their preferred habitat.

A key trend towards the end of Stage 3 is the expansion of open steppe grasslands. One model of Neanderthal extinction, put forward by Clive Finlayson, who excavated Gorham's Cave, is that the Gravettian culture was the first to exploit this environment. Finlayson argues that *Homo sapiens*, like wolves and bears, were better adapted to the pursuit hunting that led to success on the tree-less tundras of Eurasia. Pursuit hunting involves long endurance running after prey, which can eventually give up in exhaustion. In contrast, Neanderthals stuck to rivers and the forest edges, where they could continue the kind of ambush hunting they had perfected over hundreds of thousands of years. Intriguingly, the first modern humans to reach Europe, who were part of the Aurignacian culture, followed the Neanderthals in preferring warmer, more sheltered areas. As the steppes expanded, Finlayson argues, modern humans of the Gravettian culture expanded, while the Neanderthals (and modern humans of the Aurignacian culture) contracted along with their preferred habitat. This pattern may have held for the

Neanderthals in previous eras when the steppes expanded. But this time there was one crucial difference: they had company.

The Gravettian culture first appeared in the Danube corridor as early as 37,000 years ago. This was the start of a gradual cooling that, according to the Stage 3 Project, led to acute stress on the environment despite temperatures not yet reaching extremes. At this point the expansion of the grasslands was pronounced in northern Europe, but did not penetrate throughout France. Hilly areas with protected valleys, such as the Dordogne where many Neanderthal (and also Châtelperronian) sites are found, would have retained trees. By 25,000 years ago the Gravettian culture had spread to most of Europe including Britain, Spain, Portugal, Italy, Greece, much of eastern Europe and deep into Russia. At this late date the Neanderthals and the Mousterian culture were gone, and the Aurignacian culture had probably also disappeared.

What is the Gravettian culture? It is defined archaeologically by the appearance of certain tool types, a novel form of flint blade and bone points. But what really distinguished it from the Mousterian and Aurignacian cultures was that its population was much denser, with more sites in an area, that it reached a wider range of environments, notably the chilly plains of northern Europe, and that it had long-term settlements and food-storage capacity. The Gravettian is known particularly for its 'Venus' figurines, with exaggerated female forms. Unlike the meat-dependent Neanderthals of Europe, the culture involved river fishing. They had also learned to hunt the seasonal migrating animals of the steppes, in contrast to the more sedentary animals favoured by Neanderthals. It is important to note that some *Homo sapiens* using Aurignacian technology may well have been the first to develop the Gravettian industry. In other words the Gravettians may have represented an influx of new people from the Asian steppes or they may simply have been Aurignacians who developed new technologies.

Even before the first modern humans appeared in Europe, the Neanderthals had been in a stage of contraction. Why this happened is not entirely clear. It is probably a pattern that had happened many times to the Neanderthals and their ancestors over 500,000 years. They would contract to refugia during climate downturns and then re-expand. While the Aurignacian culture's arrival and subsequent volcanic eruption may have been too big a stress for the Neanderthals, the Gravettians' arrival was a game-changer. Here was a modern human group actually expanding during a cooling phase.

We turn now to the Iberian Peninsula, the area that many have argued could have been the Neanderthals' final refuge in Europe. In Asia, the pattern of extinction is less clear, and it is likely that the number of Neanderthals there was never substantial, even as they moved eastward. Iberia, as we have seen, is a special case. It has a different environment from France, Germany and the Danube corridor, and is protected somewhat by the barrier of the Pyrenees. On the Spanish side of the mountains is the Ebro River, historically a major geopolitical boundary. João Zilhão has argued that the 'Ebro frontier' approximates a line modern humans did not cross until relatively late.

By the time the Aurignacian culture gave way to the Gravettian culture in Europe, this line was breached. The question whether the Neanderthals

A 'Venus' figurine from Dolní Věstonice, Czech Republic, made of fired clay and dating from the Gravettian period.

186

lived even that long is now the subject of debate over dates. Some argue for very late Neanderthals less than 30,000 years ago in Gibraltar, others see around 35,000 years ago as closer to the last reliable date for a Neanderthal presence south of the Ebro River, while others still see no reliable evidence for either a late survival of the Neanderthals or a late arrival of modern humans in this region.

A study in 2013 tried to resolve this by re-dating bones from eleven sites in southern Iberia using ultrafiltration. Unfortunately, bone preservation in the area, as elsewhere in Mediterranean Europe, is poor, and only two sites provided material suitable for dating. These new dates turned out to be much earlier, between 50,000 and 45,000 years ago. Therefore there is a gap in reliable dates of the 40,000–30,000-year period in Iberia south of the Ebro River, which may be due to poor bone preservation alone, or to the absence of Neanderthals or modern humans or both populations from this region. Southern Iberia may have been a last refuge for the Neanderthals in Europe or may have been depleted of Neanderthals by 40,000 years ago, as were probably Crimea and the southern Balkans.

As to what possible interactions there may have been between the moderns and Neanderthals in the final days, there is a tantalizing site in Portugal. The remains of a four-year-old child were discovered in 1998 in a Gravettian rock shelter called Lagar Velho in the Lapedo Valley, about 150 kilometres (90 miles) north of Lisbon. Dating to 27,000 years ago, the so-called Lapedo child was buried with a pierced shell, the calling card of *Homo sapiens* for the previous 100,000 years, as well as with red ochre. Its body is clearly modern human. But Zilhão and the American anthropologist Erik Trinkaus argue that it also has Neanderthal traits which indicate a mixed ancestry.

Other specialists, such as Chris Stringer and Ian Tattersall, dispute this conclusion and see the Lapedo child simply as a modern human with no apparent recent Neanderthal admixture. There have been moments when this debate has been quite heated. This makes Lagar Velho the perfect place to end our story, as it reminds us that for all we have come to know about the Neanderthals, so much remains in dispute.

For Trinkaus and Zilhão the Neanderthals did not become extinct but were gradually absorbed into the intrusive modern human population. For Stringer and Tattersall, however, modern humans took over Europe from the Neanderthals, who became extinct. They have recently used the term 'leaky replacement' to account for the DNA evidence for limited interbreeding. In a sense the argument has now become one of degrees, as one side claims that *Homo sapiens* replaced Neanderthals with minor interbreeding (probably in western Asia soon after the African exodus, around

The skeleton of a child found at Lagar Velho, Portugal, was at the centre of a debate as to whether humans and Neanderthals interbred in Europe in the millennia just preceding the Neanderthals' extinction.

50,000–60,000 years ago) and the other side, citing hybrids such as the Oase Cave skulls and possibly the Lapedo child, claims more widespread population mixing. In any case, there have been so many subsequent waves of population replacement in Europe that modern Europeans have the same percentage of Neanderthal DNA as the rest of the world outside of Africa.

Why have the stakes in this argument been so high and the debate so heated? While part of it is professional pride among palaeoanthropologists who disagree, part of it must be the implications for the Neanderthals' humanity. For some it seems to be a continuation of the age-old debate over whether the Neanderthals were fully human or represent an inferior line. For example, ambivalence about Neanderthal humanity was on display in the title of *Discover* magazine's tongue-in-cheek 2011 cover story reporting the Neanderthal and Denisovan DNA admixture news, which was 'You Are Not Human'.

Our perspective, looking at the Neanderthals not as a snapshot in time but through the entire arc of their existence, is that they were on essentially the same trajectory as our species. Descended from *Homo heidelbergensis*, they became expert hunters. They developed Levallois stone tools, which required a higher degree of forward planning than the handaxes of their forebears. They survived in harsh climates by developing clothing and fire. They buried their dead and cared for the sick. They exploited a range of food resources including from the sea. They used red ochre and eagle talons, possibly for symbolic purposes, and may have painted in caves and built structures. They had language. They expanded their homeland to the Near East, Central Asia and Siberia and survived in harsh northern latitudes.

All these accomplishments put the Neanderthals on a par with *Homo sapiens* from the Eemian interglacial, 120,000 years ago. When the two species first came together in Asia, it was the Neanderthals who had the upper hand. And when modern humans with the Aurignacian industry appeared in the Neanderthal homeland of Europe, the two species seemed to coexist even as they preferred the same habitats.

Many of the Neanderthals' greatest advances occurred in Stage 3, less than 60,000 years ago. Their use of burial increased, which explains the large number of Neanderthal bones from this period. They increased the distance that they transported stone to make tools. Clive Finlayson and others argue that they were using feathers from birds such as eagles, vultures and falcons for ornamentation. They created a 'hashtag' pattern in Gorham's Cave. And there is tantalizing evidence uncovered by Zilhão in Spain that in addition to using red ochre they were experimenting with a wider range of colours and had started to perforate and decorate shells. It is not possible to say what the Neanderthals might have achieved if they had had another 100,000 years to catch up.

Only when modern humans from the Gravettian culture figured out a way to vastly increase their numbers and their range of environments and resources can we say beyond any debate that the Neanderthal population had crashed to zero. It is possible that the first wave of modern humans to enter Europe suffered the same fate. Can we say that any of these groups was less than fully human?

Human history is filled with instances of people being overwhelmed by an intruding population. In some ways this has been part of the human condition. In these cases there have often been low (and at times substantial) levels of interbreeding between native and colonizer. In historical instances where both populations are clearly the same species, it is racist to attribute one side's defeat to their biology. According to David Reich, 'The idea that the processes that were happening to the Neanderthals were happening all the time is a powerful one.'

In the Neanderthal case their biology truly was different. They were stronger, perhaps required a little more food to sustain themselves and their brains may have had smaller frontal cortices. They matured more quickly. Their brains retained the shape that our brains have at birth, and theirs were probably wired differently. Different need not mean inferior. Are we prepared to accept a different mind as a human mind? Did any of this mean they lacked the potential to thrive as we have?

One thing that all sides can agree with is that the Neanderthal way of life and body type did not survive. Whether they could have survived is something best left to our imaginations, which is where we turn in Chapter 7.

Still With Us?

'Neanderthals are among us, even now!' That's the slogan on a T-shirt we once purchased from a certain 'Saul Leviticus' at an address he dubbed Neanderthal Central Control. The species may be extinct, but there remains some truth in the slogan. Neanderthals are not entirely gone. If anything, they are becoming ever more prevalent. From a rock band to popular novels, cult movies and scientific reconstructions, Neanderthals are hard to miss. But the author of the T-shirt was probably not thinking metaphorically.

Long before the DNA investigations, people like Mr Leviticus, and indeed many university anthropologists, believed that the Neanderthal blood line at least has continued to the present day. Now that there is DNA evidence supporting the idea that Neanderthals live on within us, we'd be tempted to say that interest in the Neanderthals has increased, even though their place in our cultural lexicon was already well established. Plus, the Saul Leviticuses of this world will only be happy if a lost tribe of Neanderthals is found hiding on a mountain, probably in Asia. For most of us, Neanderthals are little more than a symbol, as we invoke their name to criticize brutish behaviour and appearance. The notion of a super-strong species of humans with strangely shaped faces who looked and acted much as we did until around 40,000 years ago is certainly unforgettable. But how close are pop culture Neanderthals to the genuine article?

Although we know a great deal of the Neanderthals' specific history, for most people they occupy roughly the same cultural space as cavemen. Their name is a synonym for primitiveness, brutality, backward thinking and generally being out of step with the times. In most cultural references, the two terms are interchangeable.

The US insurance firm GEICO created a brilliant parody of this simplistic view of our predecessors in a series of TV commercials starting in 2004. In the ads, a group of self-styled cavemen, or 'cro-maggers' (short for Cro-Magnon), who survived to the present, struggle with the negative media stereotypes about them, especially in GEICO's own non-politically correct campaign to sign up new customers: 'so easy, a caveman could do it'. Living among us as a sort of ethnic minority, the joke is that they are offended by the way we use 'caveman' as a term of derision. Though not called Neanderthals, their appearance – wide noses, heavy brow ridges, prominent teeth – is unmistakable. True Cro-Magnon men would look just like modern human men, and the humour would have been lost. This series of commercials evolved into a sit-com called 'Cavemen', whose very brief tenure on the network ABC was both a reflection of the thinness of the gag and a fitting outcome for Palaeolithic hold-overs who went extinct so long ago.

GEICO's satirical view of these popular images of Neanderthals and cavemen is also reflected in print cartoons. Like the man stranded on a desert island or the patient on a couch receiving psychoanalysis, the caveman family is a staple of cartoon humour. Most of these cartoons boil down to just two joke archetypes: in one, the caveman family does things quite primitively and, in the other, the caveman family is surprisingly sophisticated despite living in a cave. In both cases the joke plays on our preconceived image of cavemen and Neanderthals representing the opposite of civilization and, ultimately, on our deep discomfort that our advanced society arose from humble and embarrassing beginnings. We laugh not because cavemen and Neanderthals are beneath us but because they are us.

'I'm not "out of shape" – I'm *evolved!*'

Generic cavemen, often with Neanderthal features, have long been
a stock subject of jokes and cartoons, mostly uncomplimentary,
but they are typically based on outdated ideas.

This cartoon humour is a parody of the illustrated reconstructions of the everyday lives of cavemen that one finds in books like this one. In a review of iconic caveman illustrations, the anthropologist Stephanie Moser argued that many of the standard images – hunting, making tools, sitting around a fire, confronting animals, sharing a meal – are part of an established artistic tradition that actually pre-dates the archaeological discoveries and theories they claim to depict.

These illustrations often contribute to the persistence of outdated ideas. For example, virile males, often wielding clubs, are in the foreground, while women are either absent or sitting calmly in the background. Extending into the realm of the moving image, such reconstructions also perpetuate bizarre ideas about the past, such as the fact that cavemen always seem so serious and anti-social when eating or having sex. (In fact, researchers from the University of Milan argue from the study of DNA that Neanderthals, like people today, smiled as a way to indicate friendliness and win sexual partners.)

The mistaken notion that Neanderthals walked hunched over dates to a poorly reconstructed skeleton that was excavated in 1908 at La Chapelle-aux-Saints, France (see plate 13). After this skeleton was found, the French palaeontologist Marcellin Boule at the Museum of Natural History in Paris became both the first researcher to study a nearly complete Neanderthal skeleton and the first researcher to nearly completely misread a Neanderthal skeleton. Boule gave the 'Old Man' of La Chapelle-aux-Saints a decidedly ape-like, stooped posture. Now considered to have died at the age of thirty, this 'Old Man', it turns out, had bad arthritis and premature bone degeneration, and but for these infirmities would have stood as upright as anyone today. No amount of new research, tracing full bipedalism and erect posture back millions of years, seems to be sufficient to dislodge the image of the stooped caveman from our iconic vocabulary.

A key source of popular misconceptions of the Neanderthals comes from early reconstruction drawings such as this one from 1909 of the 'Old Man' (he was only 30!) of La Chapelle-aux-Saints, France.

Although Neanderthals tend to be subsumed under the generalized and mythical idea of cavemen, they do also have a particular image in popular culture. The term 'Neanderthal' is recognizably derogatory in ways no other archaic human species is. One would hardly complain of behaviour that was 'australopithecine' or 'heidelbergensian'. The meaning of accusing someone of being a Neanderthal is widely understood, even by people with little understanding of human evolution.

At times the image of a form of human ruled by animal instincts has had great popular resonance. In 1970 the band that became the legendary 10cc laid down an experimental drum track and without much thought added the words, 'I'm a Neanderthal man / You're a Neanderthal girl / Let's make Neanderthal love / In this Neanderthal world', which they sang repeatedly. The single sold 2 million copies and shot up to No. 2 in the UK charts.

The commercial exploitation of supposed Neanderthal sexual prowess continues today. A company called Primeval Labs sells Neanderthal capsules that promise to boost testosterone in middle-aged men. In Britain, Neandertal® is the name of a unisex perfume sold in bottles that resemble handaxes. Premium Neanderthal is the name of a beard-softening oil.

Perhaps on account of their unfair treatment, Neanderthals have attracted a sort of cult following by people who want to celebrate them as a noble savage, a misunderstood, thoughtful species deserving of our admiration. They are so close to being like us, yet they are the butt of our jokes. It seems that there is a widespread longing to be able to communicate with them, allowing us to atone for our sin of derision and, perhaps, for the extermination of our closest cousins.

One way we could communicate with Neanderthals would be if there were still some around today. The fantasy that living Neanderthals have somehow survived unnoticed for the past

40,000 years is part of the lore of Big Foot, also known as the Yeti, Sasquatch or Alma, depending on the part of the world where it is found. The formula is that a rural community passes down a legend of large, hairy, primitive, man-like creatures inhabiting the woods or high mountains nearby. An explorer arrives to hear first-hand accounts of these sightings, but somehow the beasts are always just out of reach.

These tales are usually fodder for novels, tabloid newspapers and TV documentaries in the tradition of spontaneous human combustion, the Loch Ness monster and haunted houses. In the 1980s an archaeologist from the University of Leicester was brave enough to attempt a relatively serious scholarly survey of the Big Foot myth, evaluating the likelihood that various sightings around the world were actually of living Neanderthals. In her book, *Still Living?* (1986), Myra Shackley dismissed these one by one, but held out hope that on a trip to Mongolia she came close to finding Neanderthals in the Altai Mountains.

Despite making such a tantalizing claim, Shackley soon left archaeology to take a post in cultural resource management, in order to study the tourism of sacred sites. With Shackley's Mongolian research unfinished, perhaps another intrepid explorer will one day continue the search and rekindle the romantic notion that our closest human cousins are simply biding their time and will rise again.

Neanderthals in fiction

In the opening scene of John Darnton's 1996 sexually charged archaeological adventure, *Neanderthal: Their Time Has Come*, a Mujahideen guerrilla takes refuge high in the Pamir Mountains of Tajikistan and is thumped on the head by a Neanderthal hungry for human brains. The notion that a surviving pocket of Neanderthals has discovered better hiding places than al-Qaeda is typical of the improbable plot devices that authors use to bring Neanderthals and modern humans together in dramatic situations.

Fictional Neanderthals have been known to hole up for centuries in mountains and caves from the Basque region of Spain to Scandinavia, Armenia and northern California. Fictional cloned Neanderthals set up their own community in Swindon, UK. And fictional Neanderthals both from our own past and from parallel universes have fallen through all manner of time machines and malfunctioning quantum computers. Only a handful of plots take place in the Palaeolithic world where modern humans and Neanderthals actually met, and while these tend to be more illuminating of how real-life Neanderthals might have lived, in our opinion there is almost nothing in fiction that approaches plausibility.

Neanderthals have appeared in the fictional works of an impressive array of writers including Darnton (a Pulitzer Prize winner), science-fiction giants H. G. Wells, Philip K. Dick, Michael Crichton, Isaac Asimov and Robert Silverberg, Nobel laureate William Golding and, inevitably, William Shatner. The bestselling Neanderthal novel is Jean M. Auel's *The Clan of the Cave Bear*, which follows the exploits of a modern human girl raised by a tribe of Neanderthals. According to the back cover of a recent edition, it has sold tens of millions of copies since its publication in 1980. In 1986 it was adapted as a movie with Daryl Hannah in the lead role (see plate 18). Auel penned five popular sequels, collectively called the 'Earth's Children' series. In addition to the air of credibility she gives her knowledge by attending conferences and basing characters on actual Neanderthal skeletons, much of Auel's success must be attributed to her deft employment of a plot turn that has been experimented with by many 'Neanderthal' authors: interspecies sex.

When Neanderthals first started appearing in fiction in the early 20th century, they were basically primitive. The seminal title of this genre is J. H. Rosny-Aîné's *La Guerre du feu* (1911), which became the hit movie *Quest for Fire* (1981). The story is based on the improbable premise that Neanderthals who could make clothes, knap stone tools and kill large predators could not rub two sticks together to make fire. When a more archaic form of human attacks a Neanderthal settlement (the Oulhamrs) and extinguishes their hearths, three members of the group set out to steal fire from a rival tribe (the Kzams, or modern humans). Speaking only in the third person and relying on their acute sense of smell, the three heroes defeat and kill a large number of scary animals and human rivals. In the film they also have sexual encounters.

More recently, fictional Neanderthals have been transformed into a gentle, gifted, morally superior people with extraordinary tracking abilities, all reminiscent of Native Americans in 19th-century literature. Jasper Fforde created a brilliant parody of this

A still from the 1981 film *Quest for Fire*, one of many visualizations of Neanderthals in the movies that have proved popular, but which reinforce erroneous stereotypes.

trend in the form of Neanderthal detective Bartholomew Stiggins in his 'Thursday Next' series. Fictional Neanderthals seem to be taking on a life of their own, becoming remarkably consistent in literature yet increasingly divergent from the archaeological view of them. On his website Fforde described the dilemma of depicting fictional Neanderthals: 'I started out by making them complete thickos, unable, in first draft, to conceive how you couldn't hijack an elevator in Swindon and order it to go to a department store in Augsberg.... Since it seems likely that neanderthals were ousted from the planet by modern man, I chose to make them unaggressive yet highly intelligent. But a different form of intelligence. They had no word for 'I' and live in perfect social order, needing no government.'

A reader who has been exposed to Neanderthals only in fiction might believe that all Neanderthals worshipped cave bears, had rigidly divided gender roles and elaborate rituals (especially surrounding burials), could track game and unfriendly modern humans with a canine-like sense of smell and possessed some sort of telepathic ability – the only question being whether they could 'see' through one another's eyes or 'share' memories. It goes without saying that none of these ideas has widespread support in the non-fiction universe. Even the plausible stereotypes are not based on any archaeological evidence.

Of all the Neanderthal paperbacks, there are two tales in particular that are thought-provoking in their depictions of possible meetings with our prehistoric kin. *Dance of the Tiger: A Novel of the Ice Age* by Björn Kurtén (1980) is a fictionalization of a theory about a possible cause of Neanderthal extinction. In the novel a modern human boy called Tiger falls in with a Neanderthal clan, marries a Neanderthal woman and seeks to avenge his father's murder. Kurtén's idea is that modern humans and Neanderthals interbred but had sterile offspring. He imagines that this could have led to the Neanderthals' extinction if the sterile hybrids became alpha males within the Neanderthal tribes but were shunned by modern humans. Kurtén was a palaeontologist, and his descriptions of Ice Age flora and fauna are rich and evocative.

The other memorable story is Isaac Asimov's 'The Ugly Little Boy' (1958) (accessible in Robert Silverberg's 1991 expansion and adaptation, *Child of Time*, and in the 1987 compilation *Neanderthals: Isaac Asimov's Wonderful Worlds of Science Fiction No. 6*; also made into a tv movie in 1977), which explores the challenges a child-care worker would face if a Neanderthal boy were brought into the present by a time machine. After a difficult start the boy adapts well to the modern world but is frustrated by his confinement in the time-machine lab, where the laws of physics somehow prevent him from leaving the room. In the end the scientists elect to send him back to his own time to an almost certain death, and his carer must figure out how she can save him. The poignant ending is a reminder of the Neanderthals' enduring humanity, contrasted with the at-times soulless march of scientific progress.

More recently, *The Last Neanderthal: A Novel* (2017), by Claire Cameron, depicts the parallel experiences of a Neanderthal girl, with the literal name of Girl, and a modern-day pregnant archaeologist who discovers Girl's remains. The book has been celebrated for how it draws upon updated theories and for its vivid depictions of Neanderthal life.

As of this writing, Anthony and Joseph Russo – the famed Russo Brothers who have co-directed popular television shows and movies, including the highest-grossing film of all time, *Avengers: Endgame* (2019) – are developing a movie also called *The Last Neanderthal*, but not based on the Cameron book.

Fauna (2020) by the Australian writer Donna Mazza imagines a near future in which evolutionary paediatricians are able to assist in fertility treatment by blending extinct DNA into developing foetuses. In the book, the narrator Stacey is helped by a character called Dr Dimitra to give birth to a Neanderthal-modern human hybrid. We never imagined when we started this project that a scientist in a work of science fiction about Neanderthal babies might be named after one of us. And yet here we are, having come full circle. Donna Mazza was inspired by an earlier edition of this book and put Dr Dimitra in her novel, and now the real Dr Dimitra is happy to say that the book is an excellent and provocative look at where the science of Neanderthal DNA might take us.

Neanderthal names

There are three basic approaches to naming Neanderthals in fiction: the primitive name (Ha), the nickname (Leviticus) and the foreigner (Stefan Antonescu).

In the primitive, grunt-sounding variety, it is common for several tribe members to have rhyming or alliterative names. For example, William Golding's last surviving Neanderthals in *The Inheritors* (1955) include Fa and Ha, Nil and Mal, Lok and Liku. Auel's clan of Neanderthals in *The Clan of the Cave Bear* is made up of females Aga, Oga, Ona, Aba, Uka, Ika, Iza, Ebra and Ovra, alongside males Dorv, Droog, Goov, Grod, Vorn, Borg, Broud, Brun, Crug, Creb and Zoug. Readers are advised to keep a crib sheet to avoid confusion. *In Quest for Fire*, Rosny-Aîné's Neanderthal names rely more heavily on vowels: Faouhm, Naoh, Ouag, Aoum.

The nicknames imply a higher degree of metaphorical thinking than the grunt names. In some cases modern humans bestow the nicknames. John Darnton's tribe in *Neanderthal* includes Lancelot, Genesis, Exodus, Leviticus (but not Deuteronomy or Numbers for some reason), Blue-Eyes, Dark-Eyes and Longface. Asimov's title character is called Timmie in 'The Ugly Little Boy'. Michael Stewart's version in *Birthright* (1990) is Adam. And then there are nicknames the Neanderthals give each other. When Robert Silverberg stretched Asimov's tale into the novel-length *Child of Time*, he gave us Silver

Cloud, Mammoth Rider, Fights Like A Lion, Stinking Musk Ox and She Who Knows. Kurtén imagined Neanderthals named after trees and flowers in *Dance of the Tiger*: Woad, Angelica, Parnassia, Torchflower, Silverbirch, Baywillow.

Foreign names imply Neanderthals who are more, or less, like us. Robert J. Sawyer's 'Neanderthal Parallax' trilogy of *Hominids* (2002), *Humans* (2003) and *Hybrids* (2003) introduces the geographically unspecific foreign-sounding names of Ponter Boddit, Adikor Huld, Delag Bowst, Jasmel Ket and Megameg Bek. Jasper Fforde's Bartholomew Stiggins is not strictly foreign, at least from our point of view (he lives in Swindon, which is not far from where we started writing this book), but his name is certainly not one you are likely to encounter in the real Anglo-Saxon world. In *The Silk Code* (1999) Paul Levinson brings us Stefan Antonescu and Max Soros, who use some genetic engineering trick in order to travel unnoticed around New York and London.

When two ethnic groups meet they tend to develop collective and often abusive names for each other, and two species would probably act in the same way. Thus the humans of Philip K. Dick's excellent *The Simulacra* (1964) refer to an ominous and backward-living population of Neanderthals as 'chuppers'. The 'wendol' are medieval cannibal Neanderthals, intentionally reminiscent of Beowulf's foe Grendel, in Michael Crichton's *Eaters of the Dead* (1976). Meanwhile Sawyer's Neanderthals call us 'Gliksins'. Kurtén's Neanderthals are 'Whites' or, less kindly, 'Trolls' to the modern human 'Blacks' of Palaeolithic Europe. Rosny-Aîné's Neanderthals call the moderns 'Thin Men'. Probably the most demeaning example is Stephen Baxter's modern humans who treat their downtrodden Neanderthal neighbours as an oppressed minority of nameless 'boneheads', a reference to their prominent brow ridges, in *Evolution: A Novel* (2002). Auel's Neanderthals are the 'Clan' who oppose the 'Others' (us), in terms echoed by Silverberg's 'People' (Neanderthals) and

Robert J. Sawyer, author of the 'Neanderthal Parallax' trilogy of novels, which feature an alternative world where Neanderthals, rather than modern humans, emerged as the dominant human species. Sawyer is better known for writing *Flash Forward*, which was adapted as a high-budget yet short-lived science-fiction TV show.

'Other Ones' (us). According to Fforde, 'Arguments of racism regarding the Neanderthals are entirely unfounded – the Neanderthal is a different species.'

Aside from the nameless beast-like Neanderthals of Baxter, these fictional names all carry the same message: that Neanderthals are just like us in that they have names for individuals and groups, and in that they have a sense of self. At the same time, the names are reminders that they are fundamentally different.

Bears, flutes and flowers

Fiction writers seem to have seized upon minority or out-dated interpretations of three cave sites in particular: Drachenloch in Switzerland is the source of the cave bear cult idea, which Auel has now etched into the public consciousness; Shanidar in Iraqi Kurdistan has given us the idea that Neanderthals were sensitive beings who decorated graves with large quantities of flowers; and Divje Babe I is a cave in Slovenia where an alleged Neanderthal flute was discovered, on a cave bear femur, forming the basis for Levinson's *The Silk Code*, a detective story featuring musical Neanderthals and Amish genetic engineers.

In the movie adaptation of *The Clan of the Cave Bear*, the human heroine is cast out of her adopted Neanderthal tribe when the tribal leader dramatically inserts a cave bear femur into a skull to mark her expulsion. This image comes directly from the excavations in the early 20th century at Drachenloch, where the excavator, Emil Bächler, interpreted this arrangement of cave bear bones, found near some Mousterian stone tools, as being evidence of Neanderthal religion. The current thinking among archaeologists is that this was just a chance configuration and there is no solid evidence that the Neanderthals arranged cave bear skulls in ritualistic ways. Even if a Neanderthal placed the femur into the skull on purpose, it is an isolated find, and there is no reason to think that it was part of a ritual, let alone a cult.

Also in *The Clan of the Cave Bear* film, there is a funeral scene in which Neanderthals decorate a corpse with a pile of flowers. As we discussed in Chapter 5, this comes from Ralph Solecki's book *Shanidar: The First Flower People* (1971), about his excavation of Shanidar Cave which had ended ten years earlier. (Auel based several of her characters on particular Neanderthal skeletons unearthed at Shanidar.) Solecki believed that pollen in the soil associated with one of the Neanderthal skeletons at the site indicated a ritual burial with flowers. Since then archaeologists have suggested another more plausible way that the pollen came to that particular patch of dirt: rodents digging holes. Common sense dictates that the evidence of a bouquet of flowers on a burial would not last over

50,000 years, but this has not stopped the spread of the notion that Neanderthals had elaborate burial rituals. For a historian of science it is almost too easy to say that the 1960s counter-culture influenced Solecki's interpretation. The latest articulated skeleton to be unearthed at Shanidar, published in 2020, was also associated with flower pollen, giving renewed attention to the flower burial hypothesis. We remain sceptical.

At Divje Babe I we have a small, newish European country with a tenuous claim on the world's oldest musical instrument. Despite evidence to the contrary, the notion that this bone is a flute has proven to be irresistible for the excavator, Ivan Turk; for Slovenia, whose Government Communication Office and National Museum promote Turk's theory; and for musicologist Bob Fink, who has kept the debate going with his self-published *The Origin of Music* (2003). The bone, which was found in a cache of other bones with no associated Neanderthal artifacts, has between two and four holes depending on how one interprets the broken parts, and the question is whether these are random tooth marks or deliberate Neanderthal drill holes. Although this discovery is relatively recent (1996, compared to 1917 for the cave bear skulls of Drachenloch and 1960 for the alleged flowers of Shanidar), it has already penetrated the world of fictional Neanderthals in a way that reaches far beyond its limited scholarly acceptance.

In a possibly unrelated development, the National Museum of Wales commissioned a jazz musician to write a seventy-five-minute Neanderthal-inspired

The Divje Babe I 'flute' was found in Slovenia and reconstructed from broken pieces, leading a small minority of researchers to believe that Neanderthals made musical instruments.

tune as background sound for its exhibition of Neanderthal teeth and handaxes. This composition went on tour around Wales in 2009, performed by singers and musicians playing admittedly fictional prehistoric stone instruments. After listening to a sample of this, we can say that it sounded like a cross between a church liturgy from a musically challenged country and a Beatles record played backwards. The US satirical TV comedy *Family Guy* poked fun at such ideas in 2005 when it depicted four Neanderthals in a cave inventing music, with their rhythmic grunts morphing into Billy Joel's 'The Longest Time'.

In addition to the popular but unsupported ideas of a Neanderthal cave bear cult, Neanderthal flower children and Neanderthal musicians, the other persistent idea in Neanderthal fiction is that of telepathy. This seems to have arisen from the fact that Neanderthal brains were on average as large as ours. The reasoning goes that if their brains were less powerful in some ways then they must have been more powerful in other ways, to account for all that Neanderthal grey matter.

Telepathy also dovetailed nicely with the outdated theory that Neanderthals were unable to speak, because it gave them another way of communicating. In the movie version of *The Clan of the Cave Bear*, the Neanderthals rely on sign language where their vocal range lets them down, but in the book this is done more with shared thinking. A Neanderthal skeleton discovered at Kebara Cave in Israel in 1983 points to strong similarities between Neanderthal and modern human hyoid bones, which help govern speech; this suggests that Neanderthals could produce a range of sounds similar to ours. And we also know from Robin Dunbar's theory of group size and the evidence that Neanderthals had the same FOXP2 gene for speech as we do, that Neanderthals almost certainly had spoken language, although it is debatable whether their speech was as complex as ours.

Neanderthals among us

In 1939 the US anthropologist Carleton Coon published *The Races of Europe*, in which he included an illustration of a Neanderthal wearing a typical 20th-century business suit and hat. While this is often cited as the root of the they-were-just-like-us school of thought, a less well-appreciated consequence is that it inspired people to test the notion that Neanderthals could walk unnoticed among us, as long as they had the right clothes. The idea of Neanderthals living in our world has become a common way of exploring just how different they were.

Every time a new museum opens with Neanderthal material, you can count on an actor showing up wearing animal skins and some sort of facial prosthetic. And TV documentaries on human evolution typically bring the past to life with the help of people demonstrating how Neanderthals might have behaved. Then there are the movies and the GEICO commercials. It is hard not to think of how the Neanderthals would have laughed if they saw our species' persistent attempts to impersonate them.

One of the rhetorical points that Coon was making with the Neanderthal-in-a-suit drawing was that it was the polar opposite of earlier reconstructions that were based on the skeleton from La Chapelle-aux-Saints. For decades Marcellin Boule's mistaken interpretation of this skeleton precluded the possibility that Neanderthals could walk unnoticed among us. The implications were that we could not have descended from them, and our ancestors would certainly not have been interested in any sexual liaisons. For this is really the subtext whenever we dress up like Neanderthals – it's a way of exploring the question of whether the Neanderthals are them or us, and whether it's conceivable that we would find them sexually attractive (which is essentially the biological definition of distinguishing them and us).

Coon effectively re-opened this question with his illustration, and in all sorts of guises we still seem to be trying to answer it. In

The archetypal 'Neanderthal-in-a-suit' drawing – Carleton Coon's illustration of 1939 – testing the question of whether a Neanderthal in contemporary garb could pass unnoticed in modern society. With Hollywood-style make-up, we are getting closer to an answer (see plate 17).

a BBC documentary called *British Isles: A Natural History* (2005), celebrity British gardening heartthrob Alan Titchmarsh donned Hollywood-style facial make-up and prosthetics to see if anyone would notice him walking down London's Oxford Street as a Neanderthal. The experiment was somewhat undone by the depth of English reserve, or just plain indifference. But the cameras managed to capture people stealing sideways glances in Titchmarsh's direction, from which we can conclude that Neanderthals would indeed stand out if they tried to blend in here.

Around the turn of the millennium British television seemed to go through a Neanderthal period, with the Titchmarsh experiment only one of a number of instances when people dressed up as Neanderthals for the cameras. The main drivers for this Neanderthal mania seem to be the advent of realistic computer animation, improvements in make-up artistry and new discoveries and theories. The BBC's *Walking with Cavemen* (2003; broadcast on the Discovery Channel in the USA), in which the Neanderthals appear in the eighth and final episode, was a logical sequel to the super-successful *Walking with Dinosaurs* (1999); it relied heavily on computer animation. As our forerunner species edged closer to the present, however, the hi-tech computer realism faded. Channel 4's *Neanderthal* (2000) also used animation in parts, notably in an impressive scene with a herd of woolly mammoths, but it was primarily about actors in Neanderthal make-up. In the USA the

History Channel's *Clash of the Cavemen* (2008) was basically an updated version of the Channel 4 series.

More recently, the BBC's *Neanderthals – Meet the Ancestors* (2018) featured Andy Serkis using technology similar to what he used to bring Tolkien's Gollum to life in *Lord of the Rings* to show us how Neanderthals might look in today's world. This same thought experiment was the basis of a feature film called *William* (2019), in which the title character is a Neanderthal brought to life through DNA technology who must navigate the modern world as the only member of his species. While this would clearly be incredibly unethical to do in real life, readers may be reassured that our knowledge of Neanderthal DNA is not nearly complete enough to allow scientists to attempt something like this.

Watching documentaries and films that feature actors playing Neanderthals, it is difficult to maintain a suspension of disbelief, which is perhaps why the light-hearted car insurance commercials and museum opening parties seem to be more effective. The best way to try to imagine the world of the Neanderthals, we believe, is not through TV shows or caveman movies, but by visiting their archaeological sites.

Neanderthal tourism

We are living in a golden age of Neanderthal tourism. In recent years new museums have sprung up at key sites and pioneering techniques have given us increasingly realistic reconstructions. The Neanderthals seem to inhabit a fertile zone at the intersection of the forces of regional development, science education and exciting new research.

Many of the major sites are now accessible to visitors, either through visitor centres or nearby museums: La Chapelle-aux-Saints and Le Moustier in France, Atapuerca in Spain, Krapina in Croatia, the Neander Valley in Germany and the Zagros Mountains in Iran.

In the Dordogne region of south-west France the Neanderthals are part of a more general package of Palaeolithic tourism, which includes much of the world's oldest and most impressive cave art. Les Eyzies promotes itself as the 'Capital of the Prehistoric World' and features an enormous statue of a stylized Neanderthal overlooking the valley below the cliffside settlement. It also opened a National Museum of Prehistory in 2004. While much of the attraction of Les Eyzies is from artistic artifacts produced by Stone Age modern humans, the town includes the site of Le Moustier, after which Mousterian stone tools were named. A few kilometres away is the Tursac Prehisto Parc, which features reconstructions of Neanderthals engaging with all manner of Ice Age beasts, and La Roque Saint-Christophe, which has similar reconstructions and is also an impressively large archaeological site.

One of the most exciting developments is also perhaps the one longest overdue. In 1996 the Neanderthal Museum opened in Mettmann, Germany, in the Neander Valley near the grotto where the earliest Neanderthal fossils to be recognized as a separate species were found. The award-winning building is home to multimedia exhibits, reconstructions, original artifacts and hands-on workshops. In the years following the museum's opening, archaeologists Ralf Schmitz and Jurgen Thissen returned to this most famous of Neanderthal sites and, astonishingly, found additional pieces of bone which fitted onto the original Neanderthal individual.

For those who prefer to visit the Neanderthals in cities, rather than travelling out to the sites where their remains were discovered, there are some excellent opportunities. In 2007 the American Museum of Natural History in New York opened the Hall of Human Origins, which features eerily realistic reconstructions by the Fossil Hominid Reconstruction and Research Team, led by Gary Sawyer and Viktor Deak. The team recreated ancient faces by building up from the skull, using clay for soft

A 'reconstructed' Neanderthal meets a young modern human at the
Neanderthal Museum in Mettmann, Germany.

tissue and lifelike urethane plastic for the skin. The Neanderthal
Campsite diorama shows a complete scene that emphasizes
Neanderthals' modern behaviours. We have since purchased our
own Neanderthal sculpture kit from the museum, so we can try
reconstructing Neanderthals at home.

In 2010 the Smithsonian Institution's National Museum of
Natural History in Washington, D.C., opened its own Hall of
Human Origins with reconstructions by John Gurche, a self-
described 'paleo-artist'. It is a positive development that artists
such as Sawyer, Deak and Gurche are becoming as well-known
as their creations. To make the point that Neanderthals are the
most human of extinct hominins, Gurche gave his reconstructed
Neanderthals stylish long hair, sometimes pinned samurai-style
at the top of the head, sometimes with blond locks flowing around
the ears like legendary rock musician Gregg Allman.

Given the importance of sites in Israel for the Neanderthal story, it is encouraging that the Natural History Museum at Tel Aviv University opened a human evolution gallery in 2018. The gallery was surprisingly contentious in that it was the first in the museum to challenge religious fundamentalist views of human origins and the age of the Earth.

Bits of Neanderthal anatomy can be found in museums throughout Europe. The National Museum of Wales in Cardiff, for example, boasts the oldest human remains in the principality in the form of nineteen Neanderthal teeth from Pontnewydd Cave, dating to 230,000 years ago. The Regional Museum of Prehistory in Halle, Germany (near Leipzig), promotes itself as Germany's oldest archaeological museum, dating to 1910. It contains the remains from Bilzingsleben along with a fine reconstruction by Paris-based artist Elisabeth Daynès of a Neanderthal posed as *The Thinker* by Auguste Rodin. Recent permanent galleries on the Palaeolithic make it an important stop on any tour of Neanderthal Europe. The Museum of Natural Sciences in Brussels, Belgium, has material from Spy, which was one of the major sites from the 19th century. There are surely others we have missed.

Perhaps 'tourism' is the word that best captures the modern experience of the Neanderthals. We feel better informed when we seek them out in museums, both in major cities and at the archaeological sites. We gape at reconstructions, trying to imagine what it would be like to meet a real Neanderthal. We laugh at their appearance in cartoons, on TV and in movies. We can dance to the song 'Neanderthal Man' by Hotlegs. We enjoy interacting with actors in Neanderthal garb, much as we would if they dressed as pirates, gladiators or vampires. But the most direct experience of the Neanderthals is one we are only now discovering – what traces of their DNA are actually doing in us.

Our Neanderthal inheritance

There is at least a little Neanderthal in all of us. This is quite a star-tling sentence to write. When we started working on this project, the prevailing view was that modern humans replaced the Neanderthals. When we wrote the first two editions, the scientific evidence was that non-Africans have a few percentage points of Neanderthal DNA mixed in with predominantly Out-of-Africa modern human DNA. Now, for the third edition, it looks like Africans have trace amounts of Neanderthal DNA as well, meaning that it is likely that all living humans have a Neanderthal somewhere in their family history.

Although the Neanderthals have been extinct for around 40,000 years, our genetic inheritance means that they are not entirely gone. While each individual today has just a small amount of Neanderthal DNA, if you take all those billions of small amounts and add them all up, according to Svante Pääbo and his team, you can reconstruct up to 40 per cent of the total Neanderthal genome. So, it is not just our archaeological endeavours that have preserved the memory of our human cousins. We collectively carry that memory around in our genes.

While our DNA binds us closer to the subject of this book, it also tells a story of distance. 'When Neanderthals and moderns mixed, they were at the edge of biological compatibility', accord-ing to David Reich, who was among the first scientists to identify Neanderthal traces in modern DNA.

Reich met with us in his lab to discuss what he described as 'the most interesting data in the world', by which he meant the Neanderthal DNA first identified by Pääbo. Reich contrasted the situation today, where all modern humans are perfectly biologically compatible, despite some populations having been in reproduc-tive isolation from others for around 70,000 years (e.g. Australians and some sub-Saharan Africans being separated from Eurasians before the Age of Discovery), with the situation between modern humans and Neanderthals, who had been separated for perhaps ten times that length of time, some 700,000 years, before they met again in Asia. This was almost, but not quite, too long for there to be fertile offspring.

Thanks to Reich and his colleagues, we know that there were fertile offspring between Neanderthals and modern humans. But we also know that the Neanderthal genes are generally not helpful to us. Superficially, people today look like the modern humans who evolved in Africa, reminding us that our modern human DNA is dominant. 'Most Neanderthal DNA has been purged from our genes', Reich told us. Some researchers suggest that several generations after the admixture took place, modern human populations had up to 10 per cent Neanderthal DNA. Much of this has been selected out, giving today's figure of around 2 per cent outside of Africa.

Of the Neanderthal DNA that remains, what is its function? Reich's team first started to answer this question in 2014 with a paper in *Nature*. At the same time, a team at the University of Washington, including Joshua Akey (now at Princeton) and Benjamin Vernot, published a similar study in *Science*. There have been subsequent studies in recent years, thanks to the gene sequencing of Neanderthal individuals from Vindija and Denisova Cave. Decoding DNA and identifying the function of particular genes is a young science, and there is much we do not know, even concerning our own regular DNA, let alone our Neanderthal DNA. By one calculation, there are some 12,000 different Neanderthal genes spread across modern populations, and we have only decoded the function of a few of them.

On the bright side of our Neanderthal inheritance are genes that may have helped our skin adapt to the harsh environments outside of Africa. Most of the rest is bad news: genes that heighten the risk of type 2 diabetes, Crohn's disease, decreased male fertility (although this has mostly been selected out) and genes that lead to a host of other unpleasant problems such as addiction, sunburn, arthritis, depression and allergies. Some have even blamed Neanderthal DNA for the modern propensity for belly fat.

Some of the unhelpful Neanderthal genes are the flip side of traits that would have been helpful in prehistory. For example, the genes for diabetes and belly fat may have been beneficial for populations coping with frequent periods of near-starvation. They have only become detrimental since the development of agriculture and the spread of the Western diet. Genes that increase blood clotting are not helpful now but might have been useful for Neanderthals and prehistoric modern humans who suffered many injuries and open wounds. Some genes related to skin and hair might have endured thanks to protections they offer against the cold, the sun in northern latitudes or perhaps are for sexual selection.

What may be most helpful for modern humans is genes related to the immune system. Since these have not been selected out, they probably bestow some as-yet unidentified protection against disease and, ironically, may even have been helpful for modern humans to out-compete the Neanderthals for planetary dominance. Gili Greenbaum at Stanford University created a computer model in 2019 that showed that when two populations meet, if one inherits a small increase in immunity from interbreeding, then that group can quickly supplant the other one.

How might modern humans have inherited increased immunity? In 2019 a team at the University of Washington led by geneticist Evan Eichler, applied evidence from copy number variations (CNVs) to the question of modern human uniqueness. CNVs occur when entire segments of DNA are copied, sometimes several times, or when some copies are deleted. These types of mutations can lead to rapid evolutionary changes. Eichler calls them the 'dark matter of the human genome' because we can only read short sequences of ancient DNA at a time and cannot always tell how many CNVs were originally there. But CNVs can be a big part of what is new in us. According to Eichler, 'The power of a duplicated gene is you make extra copies that allows evolution to tinker without interfering with the original function'.

The most common and best-known DNA mutations are single base pair substitutions, known as SNPs (single-nucleotide polymorphism). This is when one of the four nucleotides of DNA (A, C, G and T) is copied incorrectly, like a single-letter typo in a manuscript. 'Like a salt-and-pepper shaker, they are everywhere', Eichler said. SNPs are easier to see in ancient DNA but may not be as significant as CNVs for recent human evolution.

We visited Eichler to discuss our shared interest in the genetics of human evolution and autism. Eichler discovered a gene region (known as 16p11.2) that is unique to modern humans, evolved less than 300,000 years ago and appears to be related to an increased risk of autism. In terms of copy number variations, extra copies of parts of this region can lead to schizophrenia and deletions can lead to autism. He emphasized that the evolutionary changes, including an increased brain size, that make us modern are so recent that they have not reached homeostasis. This means that the prevalence of developmental disorders may be the price our species is paying for our recent, significant evolutionary changes. 'We are actually at the precipice,' he said. 'After another 5–10 million years we'll probably be just fine.'

In 2019, Eichler and his colleagues compared modern human DNA to a Denisovan, a Neanderthal from Siberia, and a Neanderthal from Croatia. They found that likely introgression from the ancients shows positive selection, probably aiding in local adaptation, diet, metabolism, immunity and cellular function. As techniques improve, the study of CNVs holds a great deal of promise in revealing what changes make us different from the Neanderthals and also what we have inherited from them.

Perhaps the biggest surprise of the evidence from our modern DNA is not the details of what we have inherited from which ancestors, but the fact that there is so much introgression from other forms of human. While our genome is predominantly modern human, it is a mosaic. Humans outside of Africa have around 2 per cent Neanderthal DNA. According to Joshua Akey, humans in Africa have around 0.3 per cent Neanderthal DNA. Melanesians and Aboriginal Australians, in addition to the 2 per cent Neanderthal DNA, have up to 5 per cent Denisovan DNA. East Asians have some as well. The Denisovans and Neanderthals interbred with each other.

To make matters more confusing, the first modern humans to leave Africa (associated with sites such as Qafzeh and Skhul or perhaps Apidima) left a trace of their DNA in the Neanderthals who later interbred with the main Out-of-Africa modern humans, tens of thousands of years later.

Recently, there have been indications of mixing from populations separated for even longer. Denisovans may have interbred with a form of Asian *Homo erectus*. And there is talk of a 'ghost lineage' of DNA in west Africans, inherited surprisingly recently from an archaic human species. There are surely more surprises to come in African DNA, which is the most diverse in the world and less frequently studied.

Beyond the few percentage points of DNA we carry from other hominins, we modern humans are all that remains in what was recently a more diverse human world. The Neanderthals are the best known, and likely most accomplished of those other varieties. We are continually learning more about them, reconstructing their lives, and imagining them in our movies and literature.

What we lack, however, is something more solemn. One behaviour that defines our humanity is the ability to mourn. Our treatment of the dead is a testament to our belief that life is more than animated matter, that our significance transcends our corporeal forms. With the Neanderthals, we do not simply have the remains of a few hundred long-deceased individuals languishing in museums, rather we have the memory of a whole species, one that lived for a few hundred thousand years, and with whom we shared a common ancestor perhaps 700,000 years ago.

There is little solace in the revelation our DNA may hold some small trace of the Neanderthals. This kind of Neanderthal survival, if you can call it that, is at best invisible and in cases such as propensity for sunburn or Crohn's disease, just a pain. What prevents us from truly missing them, from mourning their absence? Perhaps it is the distance of time. Perhaps it is their not-quite-like-us appearance.

With this book we hope we have at least taken a step towards unlocking our human feelings about our extinct cousins. By reviewing the amazing trove of knowledge that has built up in recent years, and contrasting that with the comical place they have in our popular culture, we can start to remember them for what they were and what they accomplished, rather than seeing them only through the lens of our own insecurities.

Bibliography

For the general reader

Arsuaga, Juan Luis. *The Neanderthal's Necklace: In Search of the First Thinkers*. Trans. Andy Klatt. Chichester: Wiley, 2003.

Arsuaga, Juan Luis, and Ignacio Martínez. *The Chosen Species: The Long March of Human Evolution*. Trans. Rachel Gomme. Malden, Mass., and Oxford: Blackwell Publishing, 2006 [1998].

Bordes, François. *A Tale of Two Caves*. New York and London: Harper and Row, 1972.

Dunbar, Robin. *The Human Story: A New History of Mankind's Evolution*. London: Faber and Faber, 2004.

Fagan, Brian. *Cro-Magnon: How the Ice Age Gave Birth to the First Modern Humans*. New York and London: Bloomsbury Press, 2010.

Finlayson, Clive. *The Humans Who Went Extinct: Why Neanderthals Died Out and We Survived*. Oxford and New York: Oxford University Press, 2009.

Finlayson, Clive. *The Smart Neanderthal: Cave Art, Bird Catching, and the Cognitive Revolution*. Oxford and New York: Oxford University Press, 2019.

Gamble, Clive. *Timewalkers: The Prehistory of Global Colonization*. Stroud: Alan Sutton; Cambridge, Mass.: Harvard University Press, 1993.

Gamble, Clive, John Gowlett and Robin Dunbar. *Thinking Big: How the Evolution of Social Life Shaped the Human Mind*. London and New York: Thames & Hudson, 2014

Jordan, Paul. *Neanderthal: Neanderthal Man and the Story of Human Origins*. Stroud: Sutton Publishing, 1999.

Moser, Stephanie. *Ancestral Images: The Iconography of Human Origins*. Ithaca, NY: Cornell University Press, 1998.

Pääbo, Svante. *Neanderthal Man: In Search of Lost Genomes*. New York: Basic Books, 2014.

Pitts, Michael W., and Mark Roberts. *Fairweather Eden: Life in Britain Half a Million Years Ago as Revealed by the Excavations at Boxgrove*. London: Century, 1997.

Reich, David. *Who We Are and How We Got Here: Ancient DNA and the New Science of the Human Past*. New York: Pantheon Books, 2018.

Sawyer, G. J., Victor Deak, Esteban Sarmiento and Richard Milner. *The Last Human: A Guide to Twenty-Two Species of Extinct Humans*. New Haven and London: Yale University Press, 2007.

Schrenk, Friedemann, and Stephanie Müller. *The Neanderthals*. Trans. Phyllis G. Jestice. London and New York: Routledge, 2005.

Shackley, Myra. *Still Living? Yeti, Sasquatch and the Neanderthal Enigma*. New York: Thames & Hudson, 1983; *Wildmen: Yeti, Sasquatch and the Neanderthal Enigma*. London: Thames & Hudson, 1983.

Shreeve, James. *The Neandertal*

Enigma: Solving the Mysteries of Modern Human Origins. New York: Avon Books, 1995.

Stringer, Chris. *Homo Britannicus: The Incredible Story of Human Life in Britain*. London and New York: Penguin Books, 2006.

Stringer, Chris. *The Origin of Our Species*. London and New York: Penguin Books, 2011.

Stringer, Chris, and Peter Andrews. *The Complete World of Human Evolution*. Revised edition. London and New York: Thames & Hudson, 2011.

Stringer, Chris, and Clive Gamble. *In Search of the Neanderthals: Solving the Puzzle of Human Origins*. London and New York: Thames & Hudson, 1993.

Sykes, Bryan. *The Seven Daughters of Eve: The Science That Reveals Our Genetic Ancestry*. New York: W. W. Norton & Company, 2001.

Tattersall, Ian. *The Last Neanderthal: The Rise, Success, and Mysterious Extinction of Our Closest Human Relatives*. Revised edition. New York: Westview Press, 1999.

Tattersall, Ian. *Masters of the Planet: The Search for Our Human Origins*. New York and Basingstoke: Palgrave Macmillan, 2012.

Taylor, Timothy. *The Artificial Ape: How Technology Changed the Course of Human Evolution*. New York and Basingstoke: Palgrave Macmillan, 2010.

Trinkaus, Eric, and Pat Shipman. *The Neandertals: Changing the Image of Mankind*. New York: Knopf, 1992.

Wynn, Thomas, and Frederick L. Coolidge. *How to Think Like a Neandertal*. Oxford and New York: Oxford University Press, 2012.

For the specialist reader: books

Andrefsky, William, Jr. *Lithics: Macroscopic Approaches to Analysis*. Cambridge Manuals in Archaeology. Cambridge, New York and Melbourne: Cambridge University Press, 1998.

Barham, Lawrence, and Peter Mitchell. *The First Africans: African Archaeology from the Earliest Toolmakers to Most Recent Foragers*. Cambridge, New York and Melbourne: Cambridge University Press, 2008.

Barton, Nick, Chris Stringer and Clive Finlayson (eds). *Neanderthals in Context: A Report of the 1995–98 Excavations at Gorham's and Vanguard Caves, Gibraltar*. Oxford: Oxford University School of Archaeology, 2012.

Bordes, François. *Typologie du Paléolithique ancien et moyen*. Publications de l'Institut de Préhistoire de l'Université de Bordeaux. Mémoire 1. Bordeaux: Delmas, 1961.

Camps, Marta, and Carolyn Szmidt (eds). *The Mediterranean from 50,000 to 25,000 BP: Turning Points and New Directions*. Oxford and Oakville, Conn.: Oxbow Books, 2009.

Conard, Nicholas J., and Jürgen Richter (eds). *Neanderthal Lifeways, Subsistence and Technology: One Hundred Fifty*

Years of Neanderthal Study. Heidelberg, London and New York: Springer, 2011.

Condemi, Silvana, and Gerd-Christian Weniger (eds). *Continuity and Discontinuity in the Peopling of Europe: One Hundred Fifty Years of Neanderthal Study*. Heidelberg, London and New York: Springer, 2011.

Coon, Carleton. *The Races of Europe*. New York: Macmillan, 1939.

Debénath, André, and Harold Dibble. *Handbook of Palaeolithic Typology: Volume 1, Lower and Middle Palaeolithic of Europe*. Philadelphia: University of Pennsylvania Museum, 1994.

De Mortillet, Gabriel. *Le Préhistorique: origine et antiquité de l'homme*. Paris: C. Reinwald, 1883.

Dennell, Robin. *The Palaeolithic Settlement of Asia*. Cambridge, New York and Melbourne: Cambridge University Press, 2009.

Dunbar, Robin, Clive Gamble and John Gowlett (eds). *Social Brain, Distributed Mind*. Proceedings of the British Academy 158. Oxford and New York: Oxford University Press, 2010.

Gamble, Clive. *The Palaeolithic Societies of Europe*. Cambridge, New York and Melbourne: Cambridge University Press, 1999.

Gamble, Clive. *Origins and Revolutions: Human Identity in Earliest Prehistory*. Cambridge, New York and Melbourne:

Cambridge University Press, 2007.

Inizan, Marie-Louise, Michèle Reduron-Ballinger, Hélène Roche and Jacques Tixier. *Technology and Terminology of Knapped Stone*. Nanterre: Cercle de Recherches et d'Etudes Préhistoriques, 1999.

Kuhn, Steven L. *Mousterian Lithic Technology: An Ecological Perspective*. Princeton and Chichester: Princeton University Press, 1995.

Lindenbaum, Shirley. *Kuru Sorcery: Disease and Danger in the New Guinea Highlands*. Palo Alto: Mayfield, 1979.

Mazower, Mark. *Inside Hitler's Greece: The Experience of Occupation, 1941–44*. New Haven and London: Yale University Press, 1993.

Mellars, Paul. *The Neanderthal Legacy: An Archaeological Perspective from Western Europe*. Princeton: Princeton University Press, 1996.

Pettitt, Paul. *The Palaeolithic Origins of Human Burial*. London and New York: Routledge, 2011.

Pettitt, Paul, and Mark White. *The British Palaeolithic: Human Societies at the Edge of the Pleistocene World*. London and New York: Routledge, 2012.

Roebroeks, Wil, and Clive Gamble (eds). *The Middle Palaeolithic Occupation of Europe*. Leiden: University of Leiden, 1999.

Scott, Beccy. *Becoming Neanderthals: The Earlier British Middle Palaeolithic*. Oxford and Oakville, Conn.: Oxbow Books, 2011.

Solecki, Ralph S. *Shanidar: The First Flower People*. New York: Knopf, 1971.

Stiner, Mary C. *Honor Among Thieves: A Zooarchaeological Study of Neanderthal Ecology*. Princeton and Chichester: Princeton University Press, 1994.

Stringer, Chris, Nick Barton and Clive Finlayson (eds). *Neanderthals on the Edge: Papers from a Conference Marking the 150th Anniversary of the Forbes' Quarry Discovery, Gibraltar*. Oxford: Oxbow Books, 2000.

Van Andel, Tjeerd H., and William Davies (eds). *Neanderthals and Modern Humans in the European Landscape During the Last Glaciation: Archaeological Results of the Stage 3 Project*. Cambridge: McDonald Institute for Archaeological Research, 2003.

For the specialist reader: articles

Adler, D. S. et al. Early Levallois technology and the Lower to Middle Paleolithic transition in the Southern Caucasus. *Science* 345 (2014): 1609–13.

Aiello, Leslie C., and Robin Dunbar. Neocortex size, group size, and the evolution of language. *Current Anthropology* 34 (1993): 184–93.

Aiello, Leslie C., and Peter Wheeler. The expensive-tissue hypothesis: the brain and the digestive system in human and primate evolution. *Current Anthropology* 36(2) (1995): 199–221.

Arsuaga J. L. et al. Neandertal roots: Cranial and chronological evidence from Sima de los Huesos. *Science* 344 (2014): 1358–63.

Bar-Yosef, Ofer, and Jean-Guillaume Bordes. Who were the makers of the Châtelperronian culture? *Journal of Human Evolution* 59 (2010): 586–93.

Benazzi, Stefano, et al. Early dispersal of modern humans in Europe and implications for Neanderthal behaviour. *Nature* 479 (2011): 525–29.

Benazzi, Stefano, et al. The makers of the Protoaurignacian and implications for Neandertal extinction. *Science* 348 (6236) (2015): 793–96.

Bermúdez de Castro, J. M., et al. A hominid from the Lower Pleistocene of Atapuerca, Spain: possible ancestor to Neandertals and modern humans. *Science* 276 (1997): 1392–95.

Bermúdez de Castro, José Mariá, Eudald Carbonell and Juan-Luis Arsuaga (eds). The Gran Dolina site: TD6 Aurora Stratum (Atapuerca, Burgos, Spain). *Journal of Human Evolution* 37 (special issue) (1999): 309–700.

Binford, Lewis R., and Sally R. Binford. A preliminary analysis of functional variability in the Mousterian of Levallois facies. *American Anthropologist* 68 (no. 2, part 2) (1966): 238–95.

Carbonell, Eudald, et al. Lower Pleistocene hominids and artifacts from Atapuerca-TD6 (Spain). *Science* 269 (1995): 826–30.

Carbonell, Eudald, et al. An Early Pleistocene hominin mandible from Atapuerca-TD6, Spain.

Proceedings of the National Academy of Sciences 102 (2005): 5674–78.

Carbonell, Eudald, et al. The first hominin of Europe. *Nature* 452 (2008): 465–70.

Chen, Lu, et al. Identifying and interpreting apparent Neanderthal ancestry in African individuals. *Cell* 180 (4) (2020): 677–87.

d'Errico, Francesco, and Chris Stringer. Evolution, revolution or saltation scenario for the emergence of modern cultures? *Philosophical Transactions of the Royal Society B* 366 (2011): 1060–69.

Finlayson, Clive, et al. Birds of a feather: Neanderthal exploitation of raptors and corvids. *PLoS ONE* 7(9) (2012): e45927. doi:10.1371/journal. pone.0045927.

Fu, Qiaomei et al. Genome sequence of a 45,000-year-old modern human from western Siberia. *Nature* 514 (2014): 445–49.

Fu, Qiaomei, et al. An Early Modern Human from Romania with a Recent Neanderthal Ancestor. *Nature* 524 (2015): 216–19.

Green, Richard E., et al. A draft sequence of the Neandertal genome. *Science* 328 (2010): 710–22.

Greenbaum, Gili, et al. Disease transmission and introgression can explain the long-lasting contact zone of modern humans and Neanderthals. *Nature Communications* 10 (5003) (2019).

Hardy, B. L. et al. Direct evidence of Neanderthal fibre technology and its cognitive and behavioral implications. *Scientific Reports* 10 (4889) (2020).

Hardy, Karen, et al. Neanderthal medics? Evidence for food, cooking, and medicinal plants entrapped in dental calculus. *Naturwissenschaften* 99 (2012): 617–26.

Henry, Amanda G., Alison S. Brooks and Dolores R. Piperno. Plant foods and the dietary ecology of Neanderthals and early modern humans. *Journal of Human Evolution* 69 (2014): 44–54.

Hershkovitz, Israel, et al. The earliest modern humans outside Africa. *Science* 359 (6374) (2018): 456–59.

Higham, Tom, et al. The earliest evidence for anatomically modern humans in northwestern Europe. *Nature* 479 (2011): 521–24.

Higham, Tom, et al. Testing models for the beginnings of the Aurignacian and the advent of figurative art and music: the radiocarbon chronology of Geißenklösterle. *Journal of Human Evolution* 62 (2012): 664–76.

Higham, Tom et al. The timing and spatiotemporal patterning of Neanderthal disappearance. *Nature* 512 (2014): 306–9.

Higham, Tom, and Katerina Douka. Needle in the Haystack. *Scientific American* 319 (6) (2018): 40–47.

Hoffman, D. L. et al. U-Th dating of carbonate crusts reveals Neanderthal origin of Iberian cave art. *Science* 359 (6378) (2018): 912–15.

Hsieh, PingHsun, et al. Adaptive archaic introgression of copy number variants and the discovery of previously unknown human genes. *Science* 366 (6463) (2019).

Jacobs, Guy S. et al. Multiple Deeply Divergent Denisovan Ancestries in Papuans. *Cell* 117 (4) (2019): 1010–21.

Krause, Johannes, et al. The derived FOXP2 variant of modern humans was shared with Neanderthals. *Current Biology* 17 (2007), 1908–12.

Krause, Johannes, et al. The complete mitochondrial DNA genome of an unknown hominin from southern Siberia. *Nature* 464 (2010): 894–97.

Lowe, John, et al. Volcanic ash layers illuminate the resilience of Neanderthals and early modern humans to natural hazards. *Proceedings of the National Academy of Sciences* 109 (34) (2012): 13532–37.

Mafessoni, Fabrizio, et al. A high-coverage Neanderthal genome from Chagyrskaya Cave. *Proceedings of the National Academy of Sciences* 117 (26) (2020): 15132–36.

Pearce, Eiluned, Chris Stringer and Robin Dunbar. New insights into differences in brain organization between Neanderthals and anatomically modern humans. *Philosophical Transactions of the Royal Society B* 280 (2013): 20130168.

Pike, A. W. G., et al. U-series dating of paleolithic art in 11 caves in Spain. *Science* 336 (2012): 1409–13.

Poulianos, A. Pre-sapiens man in Greece. *Current Anthropology* 22 (1981): 287–88.

Prüfer, Kay et al. The complete genome sequence of a Neanderthal from the Altai Mountains. *Nature* 505 (2014): 43–49.

Richards, Michael P., and Erik Trinkaus. Isotopic evidence for the diets of European Neanderthals and early modern humans. *Proceedings of the National Academy of Sciences* 106 (38) (2009): 16034–39.

Rodríguez-Vidal, Joaquín et al. A rock engraving made by Neanderthals in Gibraltar. *Proceedings of the National Academy of Sciences* 111 (37) (2014): 13301–06.

Roebroeks, Wil, et al. Use of red ochre by early Neandertals. *Proceedings of the National Academy of Sciences* 109 (6) (2012): 1889–94.

Rolland, Nicholas, and Harold Dibble. A new synthesis of Mousterian variability. *American Antiquity* 55 (1990): 480–99.

Rougier, Hélène, et al. Peştera cu Oase 2 and the cranial morphology of early modern Europeans. *Proceedings of the National Academy of Sciences* 104 (4) (2007): 1165–70.

Sankararaman, Sriram et al. The genomic landscape of Neanderthal ancestry in present-day humans. *Nature* 507 (2014): 354–57.

Scott, Beccy et al. A new view from

La Cotte de St Brelade, Jersey. *Antiquity* 88 (2014): 13–29.

Slon, Viviane, et al. The genome of the offspring of a Neanderthal mother and a Denisovan father. *Nature* 561 (2018): 113–16.

Steward, J. R. and Chris Stringer. Human evolution out of Africa: the role of refugia and climate change. *Science* 335 (2012): 1317–21.

Stringer, Chris, et al. Neanderthal exploitation of marine mammals in Gibraltar. *Proceedings of the National Academy of Sciences* 105 (38) (2008): 14319–24.

Vernot, Benjamin, and Joshua M. Akey. Resurrecting Surviving Neanderthal Lineages from Modern Human Genomes. *Science* 343 (6174): 1017–21.

Villa, Paola and Wil Roebroeks. Neandertal Demise: An Archaeological Analysis of the Modern Human Superiority Complex. *PLoS ONE* 9 (4) (2014): e96424. doi:10. 1371/journal. pone.0096424.

Wood, Rachel E., et al. Radiocarbon dating casts doubt on the late chronology of the Middle to Upper Palaeolithic transition in southern Iberia. *Proceedings of the National Academy of Sciences* 110 (8) (2013): 2781–86.

Zanella, Matteo, et al. Dosage analysis of the 7q11.23 Williams region identifies BAZ1B as a major human gene patterning the modern human face and underlying self-domestication. *Science Advances* 10.1126 (2019).

Zhang, Dongju, et al. Denisovan DNA in Late Pleistocene sediments from Baishiya Karst Cave on the Tibetan Plateau. *Science* 370 (6516) (2020): 584.

Zilhão, João, et al. Symbolic use of marine shells and mineral pigments by Iberian Neandertals. *Proceedings of the National Academy of Sciences* 107 (3) (2010): 1023–28.

Fiction

Asimov, Isaac, and Robert Silverberg. *Child of Time*. London, Sydney and Auckland: Pan Books, 1991.

Auel, Jean M. *The Clan of the Cave Bear* (Earth's Children, Book One). New York: Crown; London: Hodder & Stoughton, 1980.

Auel, Jean M. *The Valley of Horses* (Earth's Children, Book Two). New York: Crown, and London: Hodder and Stoughton, 1982.

Auel, Jean M. *The Mammoth Hunters* (Earth's Children, Book Three). New York: Crown; London: Hodder & Stoughton, 1985.

Auel, Jean M. *The Plains of Passage* (Earth's Children, Book Four). New York: Crown; London: Hodder & Stoughton, 1990.

Auel, Jean M. *The Shelters of Stone* (Earth's Children, Book Five). New York: Crown; London: Hodder & Stoughton, 2002.

Auel, Jean M. *The Land of Painted Caves* (Earth's Children, Book Six). New York: Crown; London: Hodder & Stoughton, 2011.

Baxter, Stephen. *Evolution: A Novel*. London: Orion, 2002; New York and Toronto: Del Rey, 2003.

Cameron, Claire. *The Last*

Neanderthal: A Novel. New York, Boston and London: Little, Brown, 2017.

Carsac, Francis. *Les Robinsons du cosmos*. Paris: Gallimard, 1955.

Crichton, Michael. *Eaters of the Dead: The Manuscript of Ibn Fadlan, Relating His Experiences with the Northmen in A.D. 922*. London: Arrow Books, 1993 [1976].

Darnton, John. *Neanderthal: Their Time Has Come*. London: Hutchinson; Sydney and Rosebank, South Africa: Random House, 1996.

Dick, Philip K. *The Simulacra*. London: Gollancz, 2004 [1964].

Fforde, Jasper. *Something Rotten*. London: Hodder & Stoughton, 2004.

Golding, William. *The Inheritors*. London and Boston: Faber and Faber, 1955.

Innes, Hammond. *Levkas Man*. London: Collins; New York: Knopf, 1971.

Kelleher, Victor. *Fire Dancer*. Ringwood, Vic., Harmondsworth and New York: Viking, 1996.

Kurtén, Björn. *Dance of the Tiger: A Novel of the Ice Age*. Berkeley, Los Angeles, and London: University of California Press, 1995 [1980].

Levinson, Paul. *The Silk Code*. New York: Tom Doherty Associates, 1999.

Mazza, Donna. *Fauna*. Sydney, Melbourne, Auckland and London: Allen & Unwin, 2020.

Rosny-Aîné, J. H. *The Quest for Fire: A Novel of Prehistoric Times*. Trans. Harold Talbott. New York: Random House, 1967.

Sawyer, Robert J. *Hominids*. Volume One of The Neanderthal Parallax. New York: Tom Doherty Associates, 2002.

Sawyer, Robert J. *Humans*. Volume Two of The Neanderthal Parallax. New York: Tom Doherty Associates, 2003.

Sawyer, Robert J. *Hybrids*. Volume Three of The Neanderthal Parallax. New York: Tom Doherty Associates, 2003.

Shatner, William. *Step Into Chaos: Quest for Tomorrow #3*. New York: Harper Voyager, 1999.

Silverberg, Robert, Martin H. Greenberg and Charles G. Waugh (eds). *Neanderthals: Isaac Asimov's Wonderful Worlds of Science Fiction #6*. New York: Signet, 1987.

Stewart, Michael. *Birthright*. New York: Harper Collins, 1990.

Sources of Illustrations

a=above; b=below; l=left; r=right

17 *Illustrated London News*, 1863; 18l Library of Congress, Washington, D. C.; 18r from Thomas Huxley, *Man's Place in Nature*, 1863; 21 S. Pietrek/Neanderthal Museum, Mettmann; 33 American Museum of Natural History, New York; 40 after Parés and Pérez-Gonzalez, The Pleistocene Site of Gran Dolina, *Journal of Human Evolution* 37, 317; 43 Javier Trueba/MSF/Science Photo Library; 46 Natural History Museum, London/Science Photo Library; 50 courtesy Nick Ashton, British Museum and Ancient Human Occupation of Britain Project/photo Peter Hoare; 55a Peter Bull Art Studio © Thames & Hudson Ltd, London; 55b after M. Roberts 1986 and J. Wymer 1982; 58 John Sibbick/Natural History Museum, London/Science Photo Library; 64 Natural History Museum, London; 78 after M. Boule, L'Homme fossile de la Chapelle-aux-Saints, *Annales de Paléontologie*, 1911–13; 81 after P. Callow and J. M. Cornford, 1986; 86l Peter Bull Art Studio © Thames & Hudson Ltd, London; 86r, 87, 93 Dimitra Papagianni; 97r Wil Roebroeks; 97l Dimitra Papagianni; 104 Brian C. Weed/Dreamstime.com; 111 The Palestine Exploration Fund, London; 112 Dimitra Papagianni; 113 Pascal Goetgheluck/Science Photo Library; 120 American Museum of Natural History, New York; 120 drawing Rob Read; 132 Michael A. Morse; 135 after E. Trinkaus, 1983; 136 Shanidar Cave, Kurdistan; 137 Anthropological Collection, Tel Aviv University; 155 Javier Trueba/MSF/Science Photo Library; 157 Antonio Rosas; 158 Ivor Karavanic; 161 after Giovanni Caselli; 169 Eberhard Karls University, Tübingen; 173 Ivor Karavanic; 176(a–c and g–h) courtesy Mircea Anghelinu/drawing Florian Dumitru; 176(d–f) Peter Bull Art Studio © Thames & Hudson Ltd, London; 178 Natural History Museum, London; 181 NASA Landsat; 186 National Museum, Prague; 188 João Zilhão, ICREA Research Professor, University of Barcelona; 194 Andrew Grossman/Science Faction/SuperStock; 195 *Illustrated London News*, 1909; 199 ICC/Cine-Trail/The Kobal Collection; 204 courtesy Robert J. Sawyer/photo Carolyn Clink; 206 Celje Regional Museum; 209 after Carleton Coon, 1939; 212 H. Neumann/Neanderthal Museum, Mettmann. Maps pp. 27, 53, 74, 107, 150–51 ML Design.

Illustrations in the colour plates
1 P. Plailly/E. Daynès/Science Photo Library; 2 Kenneth Garrett; 3 Javier Trueba/MSF/Science Photo Library; 4 Javier Trueba/MSF/Science Photo Library; 5 John Reader/Science Photo Library; 6 Kenneth Garrett; 7 Matt Pope; 8 Kenneth Garrett; 9 S. Plailly/E. Daynès/Science Photo Library; 10a courtesy Christopher Henshilwood; 10b Cave of El Castillo, Cantabria; 11 P. Plailly/E. Daynès/Science Photo Library; 12 Kenneth Garrett; 13 John Reader/Science Photo Library; 14 courtesy Antonio Rosas/photo Javier Fortea; 15 Volger Steger/Science Photo Library; 16 Antonio Rosas; 17 H. Neumann/Neanderthal Museum, Mettmann; 18 ITV Global/The Kobal Collection.

Index

Page numbers in *italics* refer to
illustrations; numbers in **bold**
refer to the colour plates.

The Dinosaurs Rediscovered
Michael J. Benton

'If you want to know how we know what we know about dinosaurs, read this book!'
Steve Brusatte, *The Sunday Times* **bestselling author of**
The Rise and Fall of the Dinosaurs

'I defy anyone who is, like me, a non-scientist to read this book and not feel a sense of wonder.'
Tom Holland, *The Guardian*

Giant sauropod dinosaur skeletons from Patagonia; dinosaurs with feathers from China; a tiny dinosaur tail suspended in Burmese amber... Remarkable new fossil finds may be the lifeblood of palaeontology, but it is advances in technologies and methods that have fostered the revolution in the field.

Michael J. Benton takes us behind the scenes on expeditions and in museum laboratories to trace the transformation of dinosaur study from its roots in natural history to the scientific discipline it is today. New technologies have revealed secrets locked in the bones in a way nobody predicted – we can now work out the colour of dinosaurs, their growth, feeding and locomotion, how they grew from egg to adult, how they sensed the world, and even whether we will ever be able to bring them back to life. Dinosaurs are still very much a part of our world.

With 120 illustrations

Homo Sapiens Rediscovered

Paul Pettitt

Who are we? How do scientists define *Homo sapiens*, and how do we differ from the extinct hominins that came before us? The latest advances in the fields of palaeoarchaeology and genetics are revolutionizing our understanding of human evolution, revealing the extraordinary story of how our ancestors adapted to unforgiving and relentlessly changing climates, leading to remarkable innovations in art, technology and society that we are only now beginning to comprehend.

Paul Pettitt takes us to the caves and rockshelters that provide evidence of our African origins and dispersals to the far reaches of Eurasia, Australasia and ultimately the Americas. Investigating these ancient sites, and the art and artefacts left behind by the hominins who passed through them, Pettitt traces the deep history of our ancestors and gives us an intimate perspective on lives as they were lived in the almost unimaginably distant past.

With *c*. 80 illustrations